I dedicate this book

SHEILA IN THE WIND

to all those of varied colour, creed, social status and occupation—some of them I still write to, but I never even knew the names of most—who without thought of reward helped me on my way.

'It is nothing,' they said, but without that 'nothing' my own efforts would not have been enough.

Author's Note

Sheila II is a 32ft gaff-rigged yawl, designed by Albert Strange and built by Dickie in 1911. She carried me alone from England to my home in New Zealand via the Mediterranean and Red Sea, past India and Ceylon, east across the Indian Ocean to Malaya, south through the Indonesian Islands to West Australia, and finally with the Roaring Forties to Tasmania and across the Tasman Sea to New Zealand. I looked after her to the best of my ability, and she looked after me when at times it was beyond my own ability to do so.

Sheila in the Wind
A story of a Lone Voyage

ADRIAN HAYTER

Lodestar Books

First published 1959 by
Hodder and Stoughton Ltd

This edition published 2020 by
Lodestar Books
71 Boveney Road, London, SE23 3NL, United Kingdom

lodestarbooks.com

Copyright © The estate of Adrian Hayter 1959

All rights reserved

A CIP catalogue record for this book
is available from the British Library

ISBN 978-1-907206-52-8

Typeset by Lodestar Books in Adobe Jenson Pro

Printed in Wales by Gomer Press Ltd

All papers used by Lodestar Books
are sourced responsibly

Republished with kind permission of Adrian Hayter's daughters: Gael Falk, Sarah van Meygaarden and Rebecca Hayter, who state: As a faithful reissue of Sheila in the Wind 1959, this book contains references and terms which are no longer considered suitable to a contemporary readership. Some attitudes expressed are also outdated by today's standards.

CONTENTS

	Author's Note	2
	Introduction by Rebecca Hayter	9
	Preface	15
	Sheila II's route from England to New Zealand	16
1	Departure from England—the first gale—navigational difficulties—exhaustion	19
2	A valuable experience—entering my first port—Gibraltar	30
3	Time in port—visitors—the Gibraltarians—a Bull Fight—departure from Gibraltar	34
4	A strong hand—a shy shark—an escort of whales—a lucky sparrow—arrive Algiers	41
5	Arab independence—a cabaret party—uneasy departure	51
6	Thoughts in the wind—arrive Bône	57
7	'Monsieur, dormez-vous?'—'Have a drink?'—Le Café du Bateau Plaisir—departure from Bône	62
8	A navigational error—arrive Malta—censorship—departure	68
9	An escapist—stormy weather—a floppy mast—arrive Derna	73
10	A New Year's Eve Ball—*Sheila*'s danger—Tobruk—on to Egypt	82
11	Arrive Port Said—Mr. MacGregor—departure from Port Said	88
12	The Suez Canal—French women—arrive Suez—a veiled proposal—clearance-out	93
13	The Gulf of Suez—planning for the Red Sea—doldrum weather—asleep among the reefs—arrive Perim Island	99

14	Another single-hander—Mike—a visit to Assab—Djibuti—arrive Aden	113
15	A hard decision—I love the Army—departure	120
16	Nearly wrecked—the need of a writer—nearly wrecked—Mukalla—departure	124
17	The south-west monsoon—exhaustion—a friend—dangerous landfall—arrive Bombay	133
18	Interlude ashore—an appendix operation—refitting—social engagements—red tape—happy departure	142
19	Two passengers—Malabar Coast—Goa—monasteries and temples—international concord	153
20	Arrive Cannanore—the lost anchor—departure	163
21	Arrive Cochin—depart Cochin—the biggest whale—arrive Colombo	169
22	The Yacht Club—the yacht California—another hard decision—departure	174
23	The rat—the doldrums—a dream—the Monsoon breaks at sea—Siamese interlude—arrive Penang	178
24	The Gurkha Officers' Club—the Home Guard—the cost of living—the Malay girl	190
25	Malacca Strait—night entry into Singapore—social engagements—departure	197
26	Planning problems—a sea of islands—the danger of rum—under armed arrest—mine-field—arrive Surabaya—the Coronation Ball	203
27	Dutch hospitality—a spoilt child—political thoughts—departure	215
28	Bali	220

29	The south-east trades—*Sheila* springs a leak—emergency rationing—a dirty bottom—communism—a hard decision	226
30	Belief—help from Kwan Yin—the condensing plant—the problem of fire—ships passed—the last sunset—underwater landing	234
31	Illegal entry—'You're in Australia now'—move to Geraldton—the use of Christian names	249
32	In the slaughter-house—the fencing contract—the shearing shed—on the wharf—social life—a harbour bar	257
33	Departure for Fremantle—the return—builder's labourer in the North-west—a 400,000 acre sheep station—hospital	271
34	Crayfishing—wrecked—out of a job	278
35	The *Trimmerwheel*—a girl—an offer for *Sheila*—departure from Geraldton	282
36	Arrive Fremantle—the Yacht Club—more crayfishing—an unlucky ship	289
37	Perth—the payment of bills—the clean handkerchief	294
38	Departure Fremantle—Planning—the Roaring Forties—a ship passed—Bass Strait—Refuge Cove	298
39	Dismal weather—the drake—meditation—departure	306
40	Adverse winds—emergency ration scale—the cyclone—land-fall—the handkerchief—the Westport Bar—home	310
	Technical Glossary	324
	Sheila II design drawings by Albert Strange	326
	Index	328

PLATES

1. Adrian around thirty years of age
2. *Sheila* in the yacht basin at Algiers
3. Anti-fouling *Sheila* as she lay against the Red House, Singapore
4. South-west Monsoon
5. I lived on Dorado whenever catchable
6. The beach where I landed in Australia and tied the dinghy to a bush
7. ...a corrugated iron affair with a falling tin chimney
8. Adrian after landing at Geraldton
9. The morning after arrival in Fremantle from Geraldton
10. Adrian on arrival at Westport, NZ
11. Sketch of Adrian by Maureen Connell for the first edition
12. *Sheila II* at her official welcome at Aurora Sailing Club, Nelson, 1956
13. Adrian sailing *Sheila* in Auckland, 1970
14. Adrian Hayter in 1970

INTRODUCTION

When Adrian Hayter set out single-handed from Lymington, England on his thirty-two-foot Albert Strange-designed yawl *Sheila II*, local betting was seven to one that he would get no further than the English Channel.

His destination was New Zealand, and the odds were definitely against him. In 1949 perhaps only eight people had sailed solo around the world, and single-handed long-distance sailing voyages were rare. Adrian, then thirty-four, was a soldier, not a sailor.

In the previous decade he had been a close observer of the Partition of India and fought as a soldier in the Second World War and the Malayan Emergency. The latter, Britain's brutal reaction to the Communist uprising of 1948, had driven his decision to sail halfway around the world, single-handed.

More than sixty years later, and in the thirtieth anniversary year of Adrian's death, Lodestar Books is republishing the story of that voyage, *Sheila in the Wind*, first published by Hodder and Stoughton in 1959. As a sailor, Adrian recounts his foray into celestial navigation, a back-street appendix operation in India, armed escort by Indonesian authorities at sea, and eating barnacles off the hull to avoid starvation. As a writer he is trying to make sense of the humanitarian disasters that brought him to this voyage. *Sheila in the Wind* is more than a report of a 13,000-mile adventure; it's a story of the human spirit.

Adrian was born to a second-generation farmer in Timaru, New Zealand in December 1914. He and his two brothers grew up farming on D'Urville Island in the outer reaches of Marlborough Sounds. Even today, D'Urville Island remains isolated and unbroken. Adrian spoke of mustering the cattle which had roamed wild all winter on horses that had also roamed wild. In bad weather, medical help could be days away so a moment's carelessness could have swift, decisive consequences. And so D'Urville Island helped to shape the man that Adrian would become.

As a boarder at Nelson College, Adrian excelled at diving, gymnastics and boxing, but his independence was at odds with team sports. He left school to work on the family farm as the Great Depression hit hard. His prospects were modest until his mother's younger sister, Adrian's godmother, made an offer: to pay all costs for Adrian, then aged seventeen, to attend Sandhurst Military College in England.

It was a long way in every sense from a D'Urville Island farm to a military career on the other side of the world, but his godmother's offer reflected family tradition. Adrian's mother and her siblings had grown up in India under British rule. For the rest of her life she recounted memories of her childhood and tales of her family's military glory. When Adrian graduated from Sandhurst three years later, she told him to join a Gurkha regiment because they are always loyal. And so, he did.

In India, Adrian continued his training as an officer in a Gurkha regiment of the British Army and was posted to the North-West Frontier. On 26 June 1940, he married Margaret Waight, known as Tigger. She had grown up in India and encouraged Adrian's growing interest in its religions. The introspective soul who had developed in the solitude of D'Urville Island began a spiritual and philosophical awakening.

When war was declared, Adrian was desperate to be in the fight and in 1942 was finally posted to Burma and the Japanese invasion. On 8 September 1944 Adrian led sixty men to take a Japanese stronghold. Although wounded, he mustered a second, crucial attack. Wounded again, he repelled a counter-attack. His persistence earned him the Military Cross; it was then Britain's second-level military honour for officers.

As Adrian wrote in his second book, *The Second Step* (Hodder and Stoughton, 1962), there was a moment in the battle when he had believed he was about to die. He had a clear vision: the perfect self that he could be and the imperfect self that he was: '... with the terrible knowledge (then too late to amend) that my discord was to go to eternity with me. And this, not death, is fear.'

That experience stayed with him through the long months of recovery, and heightened his keen awareness of cause and effect: that it is impossible to take any action, or accept any action forced upon you, without eventually facing the consequences. Ignorance is no excuse, he said, because it doesn't change the effect.

A commanding officer and life-long friend would later describe Adrian as a fine soldier with a high level of physical fitness but: 'He was difficult to lead for, as a free-ranging New Zealander, he queried most orders and disregarded others.'

Britain granted India its independence in 1947 in a rushed process that became known as the Partition. It created India and Pakistan (East and West) as two separate countries, but failed to accommodate India's complex social systems and the traditional territories of Hindus, Sikhs and Muslims. Hundreds of thousands died in a bloody civil war and fourteen million people were displaced. As Adrian wrote, 'A great terror spread over India.'

As a trained soldier in wartime, Adrian had generally supported Britain's methods; in the Partition, he didn't. He grappled with the enormity of the tragedy. In *The Second Step*, he asks: 'It became compassion, or fellow-feeling, and I could no longer view the daily reports of massacres as mere numbers but felt them as disasters happening to myself or to those I loved. Does it in fact make the slightest difference if a child is mutilated whether it is your own or another's? The tragedy of it becomes an existing fact just the same, and the desecration demands retribution from us all.'

He and Tigger had two children, a daughter Gael, born in 1943, and a son Bruce, born in 1948. The same year, Adrian's regiment was posted to Malaya after the Malayan Communist Party attempted to overthrow the colonial administration to gain independence from Britain. It ignited what the British called the Malayan Emergency.

As Adrian relates in *The Second Step*, this evolved into guerrilla warfare. The Communist bandits hid out in the mountains and depended on raiding the villages in the lowlands for food. Britain's tactics included destroying the bandits' food supply by destroying the villages' rice paddies and livestock. Villagers who resisted were shot.

Adrian was Chief of Jungle Warfare, training officers and non-commissioned officers. He didn't support Communism, but neither did he support the atrocities to fight it. He petitioned his superiors to collect rice from the villagers on a credit system and then re-supply as they needed it, to prevent it benefitting the bandits, but his efforts were futile.

He was also disillusioned with his own country. Post-war New Zealand had adopted the Welfare State. What many acclaimed as a great advance in democracy, Adrian saw as a betrayal of what he had fought to protect in war: freedom. He believed the Welfare State made people dependent on the government, and if you are dependent, he said, you are not free.

Although Adrian was made a Member of the British Empire for his work in Malaya, the Emergency continued what the Partition had started: it manifested in Adrian what would now be called depression or post-traumatic stress disorder. Perhaps it was that fierce self-reliance that drove Adrian to his own solution: to resign from the army and sail from England to New Zealand.

As he wrote in *The Second Step*: '… having experienced so much discord among people over the preceding ten years, it seemed sensible to seek harmony away from people because all the discord I had seen had been man-made.' Adrian left his wife, daughter Gael, and son Bruce in Malaya to begin his voyage from England. About six months later Bruce died suddenly of natural causes, at eighteen months old.

In England, Adrian briefly owned the forty-foot Aldous gaff-cutter *Ayesha*, but she was unsuitable for long voyages. He bought *Sheila II* and made the six-year voyage recounted in these pages. In 1953, Tigger petitioned him for divorce.

In 1959 Adrian married Dr Tamsin Lee. He promised her that his solo sailing voyages were over, but when my elder sister Sarah was a year old, the open sea called again. Adrian returned to England by ship and bought *Valkyr*, a Norwegian Folkboat, nineteen feet on the waterline. He sailed her via the Canary Islands to the Caribbean, through the North Atlantic and West Indies hurricane season, and through the Panama Canal to the Pacific Ocean and on to New Zealand.

Valkyr arrived at Nelson, New Zealand in darkness but the harbour was lit with the headlights of hundreds of cars, horns sounding, to welcome her home. Royal Akarana Yacht Club awarded its Blue Water Medal for only the second time to recognise the achievement. Adrian's third book, *Business in Great Waters*, relates this second voyage.

Nearing the age of fifty, Adrian became seamanship instructor at the Cobham Outward Bound School at Anakiwa in the Marlborough Sounds. I was born in 1963 and the following year Adrian led the wintering-over party at Scott Base, Antarctica, described in his book *The Year of the Quiet Sun*. He was awarded the Polar Medal in 1970.

Through the late sixties and early seventies he taught English in Takaka and later worked his own boat as a commercial fisherman. In 1975, increasingly dissatisfied with New Zealand's apathy against a numbing of individual freedom, he stood as independent candidate for the Tasman electorate. He stood again in 1984 for the New Zealand Party, but was not elected in either case.

Adrian would live to the age of seventy-five; his six-year voyage in *Sheila II* straddled the mid-point of his life but it didn't resolve his inner conflict. At sea he had lived as a spiritual soul and readily accepted what he called The Law: the balance of cause and effect. On land, he felt he betrayed that experience. His second marriage lasted seventeen years, but it never recovered from the hurt of him leaving for the voyage on *Valkyr*.

Through the mid-1970s to mid-80s, Adrian wrote three books: *A Man Called Peters*, *The Missing Piece* and *The Dolphin's Message*, all increasingly philosophical and challenging for most readers. He was desperate for people to understand—but defining that understanding eluded him.

Throughout the eighties he battled the issue of Accident Compensation, refusing to pay his levy because he believed it compensated people for carelessness. Every five years, he faced charges for not signing the Census, which is compulsory and therefore to Adrian undemocratic. He went into the bush for six weeks to avoid being sent to jail.

As he neared seventy, the years under tropical suns and heavy smoking manifested what I describe as Adrian's last great adventure: cancer. It paralysed and disfigured one side of his face. He resisted conventional medicine, unable to believe in its healing beyond a physical level, and lived alone in his caravan at the Wairoa Gorge, near Nelson, always writing.

'The success or failure of fighting cancer,' he wrote, 'is not necessarily determined by death.'

People often describe Adrian as an eccentric. For others, he was a mentor, an inspiration. A week before he died at home on 14 June 1990, he said: 'I feel an overwhelming gratitude to life itself, and to share its beauty with another is now my awareness of God.'

He had finally found peace.

Rebecca Hayter
Golden Bay, New Zealand
May 2020

PREFACE

The greatest difficulty in writing a book such as this is to reconcile the two widely conflicting aspects of it, aspects which in truth cannot be separated. This story of sailing a small ship over thousands of miles of open ocean could be written purely as an adventure story, except that the ship and the sea were only a means to an end. I used these in my search for something greater than either of them, and those thousands of miles have given me a clearer idea of what it is I seek and why it eludes me.

So perhaps this story should be written more as a philosophy but after much thought I know that the two aspects should be welded into one, as indeed they were in fact. Each had its effect for better or for worse upon the other, and neither would have been the same if left to stand alone; nor is it possible in a true story that either should do so.

But can such a story be written? The reader seeking adventure is not interested in the author's amateur ramblings into philosophy, while the thinker seeking to solve the basic problem which confronts us all, whether we are conscious of it or not, is not interested in handling a ship or messing about in boats. It is possible that literary technique could soften the impact of these two aspects, harmonise them into a smoothly readable book, but that would be the falsest tale of all. For this impact in actuality is often harsh, and our reactions to it affect what we see around us and what we ourselves become.

And so this book is merely a true story of the physical adventure, a truth that is incomplete, although it does contain some thoughts that will make strange reading from an armchair; but I am not reporting thoughts from an armchair. And some day perhaps I shall write the rest of the story, the other aspect of it, and so complete its truth and fulfil what I deeply feel to be an obligation, because a half-truth may do more damage than a lie.

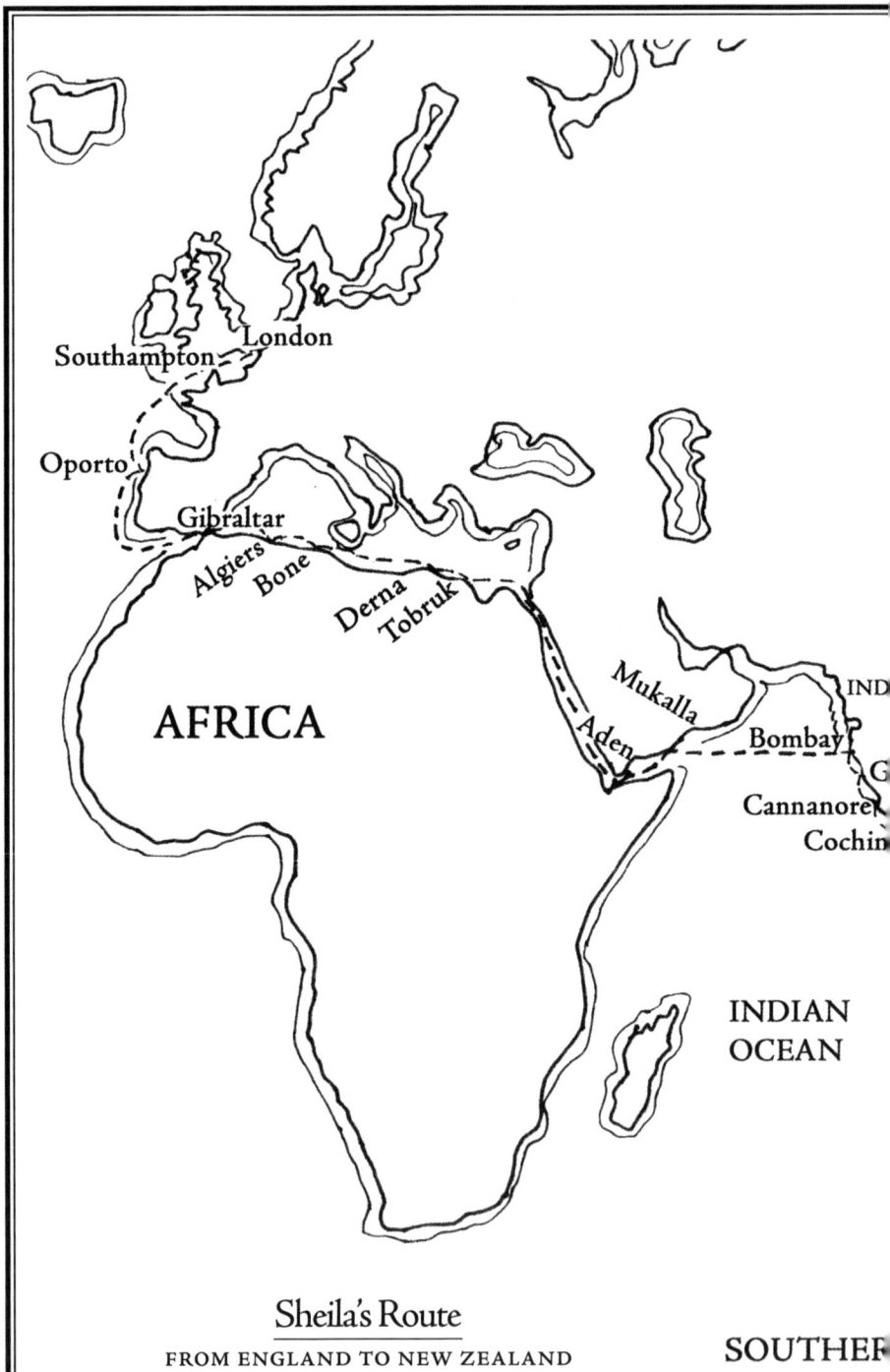

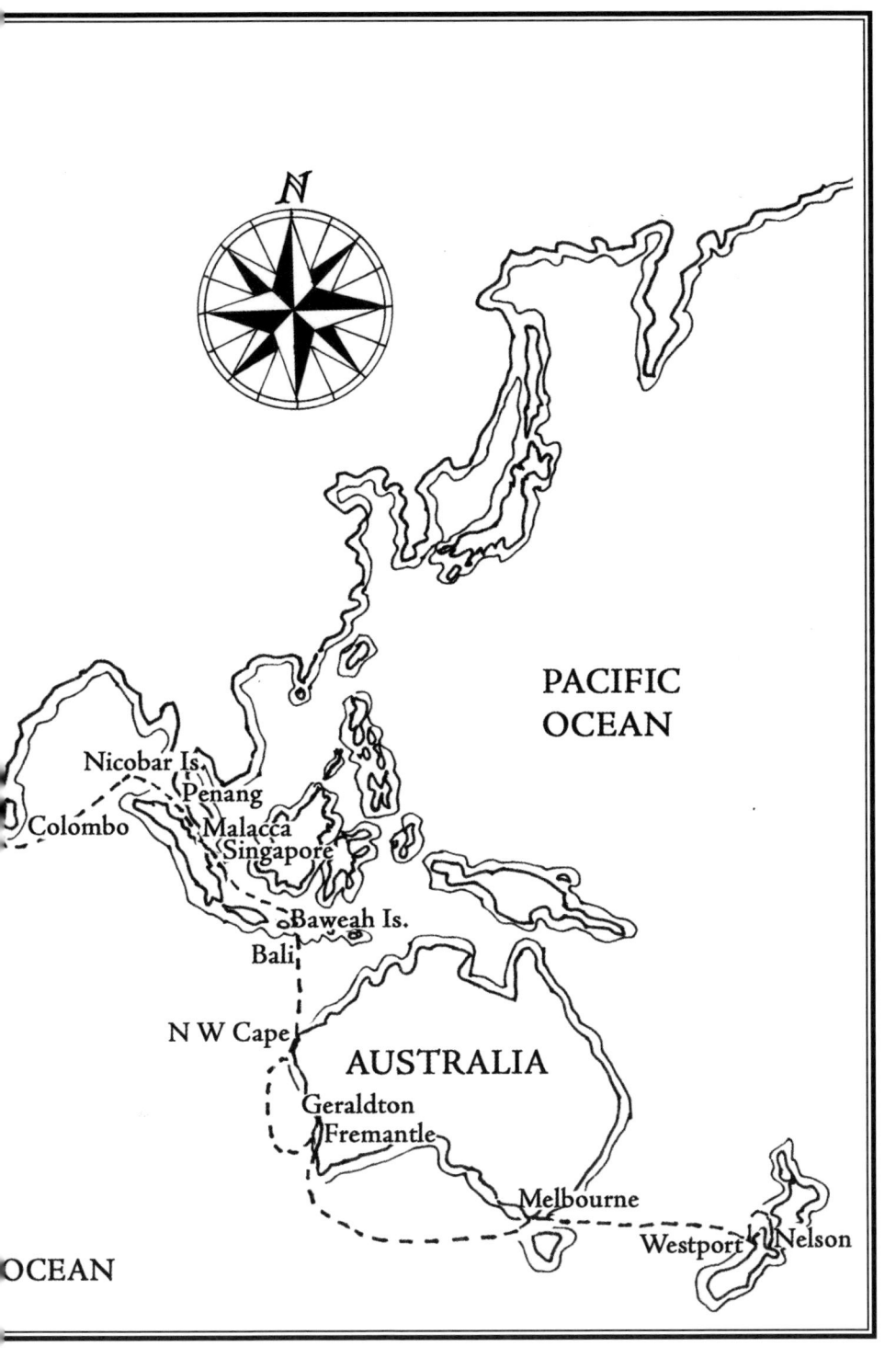

I

*Departure from England—the first gale—
navigational difficulties—exhaustion*

Some London papers printed the news that I was about to sail, and this was followed over the next week by a flood of more than a hundred letters from people who wanted to come with me. Few had any experience of the sea, and fewer still had any money to offer towards expenses.

Extracts from some of those letters will give a general impression of their variety, and how difficult all were to answer. One read: 'I am twenty-five years old, physically strong and pretty clued up—already I've made and lost three fortunes in different parts of the world, so you can see I'm the adventurous kind and am just the companion for you. At present I have no money but what's money anyway—all I ask is my keep and together we'll sail the world, overcoming all difficulties.' At the other extreme was, 'I'm now fifty but still fit and willing to work. Your venture is something I've always wanted to do, but have been prevented by obligations which could not be neglected. I don't mind if I die on this trip so won't be afraid, but will you please give me this last opportunity to do something I've wanted all my life? If you can't take me, know that my prayers will be for you. P.S. I have a little money and it is all yours.'

There were about ten letters from women and girls, some included photographs, three bore no address or signature. One was very confident: 'I am an adventurous person, more like a boy than a girl, and love to do crazy things. I'm never afraid and will do everything you tell me to (love, honour and obey!). Take me.' An anonymous letter said simply, 'I have just seen your photo and would love to come with you. You look so lonely—I too am lonely. I will pray for you and hope to meet you someday'. The dreams people have, the escapes they seek, and the dreadful loneliness some feel: I knew them all.

All those letters (except the anonymous ones) requested early replies. Many I wanted to answer, not only to say 'no' as gently as possible, but to thank them also. At this time last-minute arrangements demanded full attention, and so I drafted a reply which would do for them all, and had copies typed and addressed to the writers. That letter must have seemed very unappreciative, even callous, to some; should any of them read this I hope they will understand and forgive.

The last night I went along to the Ship Inn. The Ship, only yards from my mooring in the river, had been a second home to me over the past year. Paddy and Peggy (mine host and his wife) had been very kind to me, from introducing me to other guests so I would not become too much of a recluse, to putting me to bed and looking after me when I was ill.

This was the night—all worries were over and payments were made and I was free to sail, my drinks were on the house, many friends were present; the stage was set for a hilarious celebration. I hardly said a word the whole evening. Terrified! It is strange how you can plan, work and sacrifice towards a thing for months, and yet only when it is within your grasp do you fully realise the enormity of what you're doing. I suddenly realised I was about to emulate what people read of with awe—Slocum for instance. Taking another whisky I asked myself who the hell was I to emulate Slocum? A telegram handed to me during the evening did not help. It read, 'Adrian don't do it.' Then someone else told me that the local betting was seven to one against my ever clearing the English Channel. Thank God he told me. I left the party and went to bed, miserable!

Next morning, Saturday, 12 August, 1950, I slipped quietly away from the Berthon Boat Company, motored past the Royal Lymington Yacht Club unnoticed, much to my relief, turned off the engine and put on sail with two reefs down. The first thrill of this venture came as *Sheila* lifted to the swell and heeled to the first gust by Jack-in-the-Basket at the mouth of the river. The 1400 miles to Gibraltar seemed like a trip to the moon.

All that night I beat down Channel against a gusty sou'-wester. It rained a lot, and I was twice nearly run down by steamers coming up from astern. It was so unusual sitting alone in the darkness of the exposed cockpit, huddled against the cold and wet and partly bemused by the one tiny compass light, that I felt I was a mere solitary atom in a great empty void and forgot about shipping. The noise of their engines made me turn around to see the red and green lights, with the masthead lights in line.

I slept for an hour the next day but stayed at the helm all that night; there was too much shipping about to sleep safely, apart from the fact that it takes time to get used to sleeping alone at sea, leaving the ship to look after herself. That night the mizzen cross-trees broke in a squall and next morning, being very tired, I decided to go into Teignmouth where there were friends.

Being somewhat unfavourably impressed with my first taste of single-handed seafaring, after an agony of thought I wired a friend that a berth was his if he still wanted to come. The agony of thought—it is amazing how difficult it is to know your own mind, and amazing too how you can argue yourself into anything. All the arguments I had rejected for taking someone with me I now enlisted as being obvious common sense. However deeply honest I tried to be with myself, I just did not know whether I was giving in or being sensible. So I sent that wire, leaving the decision to some power knowing better than I. Next day a wire arrived saying he could not come, and the day after a letter in fuller explanation. All the arguments I had used over the last six months for not taking him and which he had rejected, he now quoted as reasons for not being able to come. 'But,' he added, 'if you're stuck and feel you can't go on alone, I'll sell out and come with you.' That was a generous remark, so I wrote back and said of course I could go on alone—and only then felt that I really would.

On the 19th the Met. Office told me the weather was uncertain until the outcome was decided between a depression over Iceland and an anti-cyclone over Central Europe. The issue had still not been decided on the 20th or even the 21st, and so I sailed that morning. An extract from my diary written after sailing tells of that better than my memory now:

> It is a strange feeling starting off on a trip like this—God knows how or where it will end. I can still see the white houses and hotels along the front, the red cliffs, the steep forehead of the Ness, and behind are the green hills of Devon. It is a beautiful day, a light southerly breeze and a warm sun. As I was getting the dinghy on board before slipping the mooring a young chap rowed alongside and asked to come with me. A launch caught me up going down the Teign, a photographer on board, and two reporters (one a woman) sat below while the photographer did his stuff. Later they remarked on my books—Charles Morgan, Huxley, Plato, Shaw, Tagore, and so on. The man asked, 'Are you in search of a new philosophy?' and the woman, 'Is there a girl at the other end?' They were nice.
>
> Now the launches and speed boats have gone, I've had lunch, cleaned up, and am enjoying the sun, looking back on the England I've grown to love so much and wondering if I'll ever see it again.

During the first two days out of Teignmouth, the wind was erratic and I was continually changing jibs, reefing and unreefing, and snatching odd hours of sleep during daylight because of the fear of being run down at night. The glass dropped slowly and steadily, assuring me by the evening of the 23rd that I was shortly going to experience my first gale.

The gale developed next morning so I hove to and for the first time changed to storm canvas. I had practised this in port, but it was a very different matter at sea. Before beginning I read it up again in the text book, and the only thing that this forgot to mention was the most difficult part of the whole procedure—how to stay on board while doing it. In such jobs, when both hands are used, you hang on with your legs, the crook of an arm, the point of your shoulder, and often use your teeth as an extra hand.

I had fervently hoped that I would not meet a gale until I had gained some experience of the sea, but running straight into one was the best thing that could have happened. I sat for hours in the open cockpit as *Sheila* lay hove to under storm canvas, because being down below under closed hatches frightened me out of my wits. Whenever a breaker crashed on board it sounded as if the masts had gone, planking and timbers smashed to pieces and immediate sinking was inevitable. I'd rush to the hatch, tear it open expecting to see chaos, to view *Sheila* as if nothing had happened; and nor had it. Later, common sense told me that if a breaker did smash in the decks I would know about it without the need to look above.

Also when below there was a feeling of neglecting the ship, of not being ready for emergency at the time it was most likely to happen. For the next forty-eight hours of that storm I hardly left the cockpit, and during that time I did nothing for the simple reason there was nothing to do. The sails were set hove to, reduced to their smallest size, the helm was lashed, and I could do no more but leave it to the gods. I was wearing myself out quite unnecessarily and very stupidly.

This gale also gave me the essential confidence that *Sheila* could look after herself in big seas. Time and again a huge steep sea, a veritable wall of water, seemed to fill the sky above and I'd think, 'She can't get over this one,' and *Sheila* rose gracefully, somehow slipped around the jagged crest, and sank easily into the next trough.

I had no idea whether or not *Sheila* would sail in that weather under storm canvas, and there was only one way to find out; it was something I had to know for the safety of future navigation, when storms might catch me in congested waters or close in to a lee shore. That sail proved to be one of those thrilling, fantastic experiences you never forget—like your first solo loop in an aeroplane, your first fast run on skis, anything in which temerity and doubt are banished to leave a whole new field of adventure open to you. And adventure in this light is freedom.

I had splashed through rough seas in a motor launch around the coast of New Zealand, but here in the Channel there were noticeable periods as we climbed the hills. From the top was a view to all horizons, of the long lines of rolling white-topped seas and the deep watery valleys between them. Sometimes *Sheila* slid swiftly down the steep slope into these, at others her bow hovered before we fell headlong down a watery cliff to the valley floor below, only to land, it seemed, on rubber cushions. *Sheila* said, 'You look after me, my boy, and I'll look after you.' And that was fair enough; it was like being back with Gurkha troops.

Over the next two days I got very tired. I only slept by day in hour snatches and this robs sleep of much value, delaying fatigue enough to carry on for a few more hours but building up no reserve of energy. The long night hours in the exposed cockpit were bitterly cold; there was much spray, at times breakers smothered the ship, and heavy belts of rain added to the misery.

In these days too I overworked myself through sheer inexperience. I'd think the wind had eased, and change from storm canvas to the closely reefed mainsail only to overburden *Sheila* unmercifully, and have the weary process of changing back again. I had little knowledge of what stresses and strains *Sheila*, the rigging and the sails could take, which also led to much anxiety and needless worry.

(Prior to this voyage I had never been out of sight of land in a small boat. Since arriving in England a year before, I had sailed mainly around the Solent, putting the theory of the text books into practice; before that I had only once before stepped on board a yacht when a Christmas party had called at my home on D'Urville Island, New Zealand. We youngsters had gone on board to 'see inside'—beyond this and brief outings in sailing dinghies my yachting experience was nil.)

I wrote, 'I have to stay awake all night in case of shipping. It's bitterly cold and I'm not really enjoying this very much. There are several ports on both the French and English coasts within easy reach, but I know if I go into a port without beating this ruddy gale I'll never leave land again.'

Then came the first Sunday, and a blessed day indeed, a beautiful sunny day with a gentle wind, and the seas died to let *Sheila* sail herself happily towards Ushant. I brought out the soaking mass of clothing from below to dry on deck, and cooked myself a curry lunch. When stripped to change into dry clothes I was amazed to find that my body was black and blue from head to foot, and in the most unexpected places, from being knocked around during the rough weather.

That night I put lights in the rigging and slept dead to the world until eight o'clock next morning, and it was a beautiful morning. This was the first real sleep I'd had since leaving, and there was something wonderful and strangely exciting in waking to such a lovely day out of sight of land, of ships, of people. And that new world was me.

With dusk a strong westerly came in from the Atlantic, driving *Sheila* hard across the increasing seas to fly past the winking light of Ushant—spray flying, flashes of blue lightning and giant rolls of thunder, and sometimes the moon shone briefly between the towering white columns of the thunder-clouds to show the upward curve of the white sails, the wildness of the tossing seas and their racing black shadows. Ushant died astern and home waters were behind me.

The westerly continued all next day carrying huge seas before it, but lacking the viciousness of those in the Channel. I hove to for a couple of hours' sleep during the day, and sailed all through that night into the next dawn. I wanted to get out of the notorious Bay as soon as I could, and there were another 300 miles to go. I slept for an hour after dawn but by noon was steering so badly, half-dazed with fatigue, that I feared that if anything went wrong I'd be in no fit state to put it right. And once my head nodded forward as a sharp sea lifted *Sheila*'s quarter (the corner of the stern) catapulting me right out of the cockpit to land impaled on the lee stanchion. That was another lesson, and thereafter when tired in rough weather I always tied myself in with a lashing round the waist. I hove to and slept again.

Two hours later I shot out of my bunk and through the hatch with the realisation that a loud hooting in my dreams was real. A big tanker (the *Shelldrake*) was rolling heavily beside me, the skipper peering over the end of the bridge right into my hatchway, and I heard his laugh as I appeared frowsy and alarmed. They must have had a discussion before leaving their course to investigate this lone yacht apparently unattended. I regretted the intrusion, but was grateful to a ship which troubled to ascertain that help was not needed.

The weather looked black to the south-west, and as the day grew old wind came. I tore the mainsail badly getting it up, so changed to storm canvas and let *Sheila* sail herself away from the shipping lane so I could get more sleep during the night. I was too tired to mend the sail then.

Next day was fine with a gentle breeze, so I mended the tear and let *Sheila* sail herself to the south while I did all those things I ought to have done the day before. It is very damaging to morale to leave those things undone that you know ought to be done—'and there is no health in us'. How exactly that describes the feeling, of apprehension, of 'butterflies in the tummy'. It is also (I have found) the basic cause of bad temper, even when alone. The solution is obvious.

I ran the engine to charge the batteries, filled respective containers with meths., kerosene and petrol from the bulk stocks, cleaned the navigation lights and trimmed the wicks, tightened the port shrouds, cut out some chafe in the starboard runner (an extra back-stay for the main mast) and put in a long splice (copied from the book!). I gargled with antiseptic as I'd had a sore throat for a couple of days, washed up the dirty dishes and put them away, scrubbed out the galley, burnished the draining-board and de-greased the sink. Clothes and blankets were aired and re-stowed, spare sails neatly rolled and stowed, cabin tidied, ropes above deck coiled neatly, and the bilges pumped dry. In fact *Sheila* was as shipshape as the day she left, and after a shave I felt even better than I myself had on that momentous day.

Cape Finisterre was somewhere about 120 miles to the south, and as I'd now been four days out of sight of land I took a sun sight, the first ever. Before sailing from England I had started a correspondence course on celestial navigation, and gone to retired Master Mariners for instruction, but all these went far too much into detail for me. I can only understand theory when I put it into practice, and so having bought the necessary nautical tables and almanac I threw the papers on board for further study when the time came. The risk of such a system I offset by planning to keep about 100 miles offshore after sighting Spain, to allow for errors while under instruction. It would be impossible to miss Gibraltar, wedged in between two great continents.

The position line obtained from this sight ran near my dead reckoning (DR) position, and as clouds were coming up to obscure the sun I headed south-east to the shipping lane, which runs near enough in a straight line from Finisterre to Ushant. This line I crossed with the position line, and from the intersection laid a course for Finisterre. This showed up two days later.

The next four days and nights were spent mostly encased in a thick fog. There was a big lazy swell and little wind—conditions which give a sailor a glimpse of hell. If you leave up sail the rolling of the ship slams the sails and spars from side to side, jerking the blocks, wearing the sails, and chafing the gear; if you take sail down you understand the meaning of the term 'rolling her guts out'. The almost continual hooting of fog-horns on that busy shipping lane did not help to ease the strain. My fog-horn had made a raucous blare in harbour, but in that great white loneliness it sounded like a toy flute. Sometimes ships passed so close, but invisible, that their breaking wash smothered the side-decks.

As each day and night passed with little change in conditions I reached a stage of frustration which nearly drove me mad, and the engine was not much help because I wanted to save at least ten gallons of fuel in case of need in the confines of Gibraltar Strait. The noise or movement made sleep uncertain, and any breath of wind I used to the full, at whatever hour of the day or night it came.

Then I began to feel really ill, and searched desperately for the cause. A few days previously the tea jar had been broken, and rather than throw the tea away I'd sifted the tiny splinters of glass from each lot before use; perhaps some had been missed. Or perhaps some tinned food had been faulty, or perhaps the remains of some dread tropical fever still lurked in my blood. It felt like all of these—a splitting headache, extreme lassitude and depression, and a wracked tummy which in five minutes miraculously converted all solids into liquid. So I took no food and increased my weakness further.

Yet the ship had to be worked and I must have made progress which my diary is too vague to show. My log tells me that I took sights two days later, putting me forty miles off the Spanish Coast, which was safe enough.

Before going below to make a cup of cocoa at dusk I took a routine glance around the horizon, and to my horror saw a mass looming out of the fading light broad on the starboard bow. Compass bearings and the Pilot Book told me that I was inside a crescent shaped chain of islands composed of Berlenga and the Farilhoes, about forty miles north of Lisbon. There was no need to panic but that is exactly what I did.

I clawed down the sails and started the engine, turned northwest and tried to motor out against a steep sea and freshening wind. This was a useless waste of my limited energy and the engine's fuel, because *Sheila* can deal with such conditions far more efficiently under sail than under power. So I stopped the engine, hauled up sail, put in two reefs, and went on to the port tack. Sometime during the blackness of that night we passed the outer-most rock, when and by how much I don't know. Short tacks would have eliminated all risk; I had effected only two and then left *Sheila* to sail herself, taking a chance on the wind not changing, and going below to the dimly-lit cabin too tired and ill to do more.

The next day brought the crisis. I was standing in the cockpit leaning against the coaming when my knees buckled. My mind was strangely clear and detached as it watched my body crumple and sink into a huddle on the cockpit floor.

I came to some time later, noted that there was wind and that for some reason the mainsail was down. I dragged myself forward to the halyards, not daring to stand upright for fear of falling overboard, but I had not the strength to pull the sail more than halfway up.

It is hard to describe that humiliation now and so I quote from my diary (written up later that evening):

> How I wept today—I vaguely remember sliding on to the floor of the cockpit, my head falling on to my doubled-up knees and sobbing my heart out; I tried to stop and just couldn't. All my stupidities seemed too big and yet unended, just as people who knew more of the sea, more of the Army, more of life, had predicted. My navigation is all to hell, I know nothing of handling a

ship and have killed myself trying, my capital's gone, my career's gone—oh brother, are you a failure?

I was laid completely bare and the whole world seemed to be in the watery sun looking down at me, so I went below as a wounded animal goes to hide in a cave, flinging myself on the berth as I thought to die. Then fatigue mercifully overcame the pain all through my body and I lost consciousness.

2

A valuable experience—entering my first port—Gibraltar

A round patch of light moved to and fro over the other berth as *Sheila* rolled quietly to the gentle swell. It was nearing noon, and my first realisation was that all pain had left me; I felt refreshed and at peace—and hungry. This told me that I'd only been a bit tired, nothing seriously wrong, and the black despair of failure and doubt had gone with the pain.

I fried up some bully beef and onions; it was good, but unfortunately fried bully tastes just like fried bully. At first I wolfed it down, but as soon as the first sharpness of hunger was gone, I started to put aside those horrid-looking bits of gristle and the flabby slivers of thick white skin, which make up heaven knows what part of an ox. This meal was a celebration, so ran to two courses, peaches and cream (condensed milk) with a cup of coffee after. Then I went back to bed.

I woke in the evening. After supper while the washing-up water heated I sat on the foredeck with a mug of coffee, watching the most beautiful sunset I've ever seen. In its beauty lay forgiveness and the peace that comes with it.

This experience of illness was invaluable merely from a sailor's point of view, and indeed was part of the very careful plan I had made before the trip began—to drive myself to the limit in order to find out just what I could stand under those conditions. It is essential to know this for safe navigation, in exactly the same way as if you have a car with a broken petrol gauge. Before you know how far you can go, what stretches of the journey can be completed without replenishing, where you must replenish and so on, you must drive that car from a full tank until it is quite empty. For the same reason I drove myself until I actually passed out from sheer exhaustion, although it was not recognised as such at the time.

After that I knew not only roughly how long I could expect to last under those conditions, but even better I now knew the warning signs of approaching exhaustion and its inevitable inefficiency, signs which came in time to anticipate and plan to avoid that final state of collapse, and to put my ship in a position of safety before that state did arrive. You can sleep as much as you like far out at sea, but near land or shipping it is stupidity.

There must have been fair winds during that week because I was then in the area of the Portuguese Trades, which blow from the north-east. My DR put me forty miles north of Cape St. Vincent, the south-west corner of Spain and the last to round before the home run to Gibraltar. Land was beyond the horizon to the east, and at noon I took infinite trouble in ideal conditions with my first meridian altitude sight, which gives latitude. The line put me forty miles south of my DR, exactly opposite the Cape. I checked and re-checked the simple working, not daring to believe the gift of forty miles, and turned east to pick up the coast to make sure. In the late afternoon the haze lifted to show the exact original of the sketch of the Cape displayed in the Pilot Book. That night I passed south of the light, deliriously happy to be on the home stretch.

Four days later, gigantic mountains appeared out of the dawn on either hand, their height exaggerated by a thick white mist which hid their bases and all other signs of land. They seemed to float, celestial, like mountains in a Chinese painting.

Soon a school of whales ambled round *Sheila*, lying lazily on the surface very close together in pairs, like contented married couples taking their morning tea. Some came very close astern and one detached itself to swim right alongside just under the surface. It looked bulky from the side, but from directly above the heavy shoulders sloped into the surprisingly long slim tail, which in turn flowed outwards into the beautifully-shaped flukes. Each of us I think was quite new to the other, and regarded each other with the frankest interest overcoming any suspicion of ill intent. Suddenly doubt passed like a current between us and he turned away with a great swirl, the flukes of his tail missing *Sheila*'s side by inches.

At 10 a.m. the haze lifted and gave me my first glimpse of the Rock, my goal for the past four weeks, which had been the hardest four weeks I have ever known. At first a wave of wild elation swept over me, then it changed to one of quiet content.

An hour before arrival I hove to and prepared for entry into port. Above decks I let fly the club burgee from the mast head, the Red Ensign on the stern, *Sheila*'s registration letters with Lloyd's (E I C D) from the port signal halyard to the main cross-trees, and the yellow 'Q' flag from the starboard, requesting *pratique*.

Before entering a strange port it is not possible to know whether you will tie up alongside a wharf, lie to an anchor, pick up a mooring, or lie to an anchor with a stern line ashore, and so you have to be prepared for any of these. I unlashed the two anchors and placed them ready to let go, unlashed the boat hook and placed it handy to the cockpit, laid fenders ready on either side, tested the engine, and laid a warp bow and stern ready coiled to throw, and three lighter ropes amidships for use as springs if needed. The ship's papers, the Health and Customs clearance-out papers from England, and my passport were laid ready for inspection on the saloon table.

I then tidied up down below (I was too excited to wash-up and so hid the dirty dishes), had a shave, combed my hair, changed into a clean shirt, and took another quick look at the Pilot Book regarding entry regulations for Gibraltar. I then headed for the South Mole to meet the port boarding launch and report, and to ask for berthing instructions. This was my first ever foreign port in *Sheila II* and was thrilling; a thrill, I was to discover, which never faded even after many destinations had been reached and passed.

The port authorities gave me clearance-in and directed me to an anchorage opposite the causeway between Gibraltar and Spain, where I let go in three fathoms. The police launch came alongside and told me not to leave *Sheila* unattended, because Spanish fishermen would probably come out and strip her of ropes, blocks, anchors, and anything movable. Similar warnings were to be given by the authorities in every port I entered between there and New Zealand, but only once was anything stolen from *Sheila*, and that was by dock workers not by fishermen.

I was longing to get ashore, to see people again, to have a good meal, send cables home, collect mail, but more than anything to buy some cigarettes. I had smoked from forty to sixty a day for years and had often tried to give them up, enduring agony and sometimes getting as far as lunch-time without touching one. So I had deliberately sailed from England without any in a desperate attempt to break the habit, working on the theory that if I couldn't get any I wouldn't smoke any. As the shops lit up that night ashore with stacks of cigarettes so nearly available, the joy of smoking again was eagerly anticipated.

After supper that evening I sat on deck as *Sheila* lay quietly to her anchor, watching the lights ashore, the car lights climbing the higher roads above the town, hearing the hoot of taxis, and music from a nearby ship—my radio had been out of action for the last three weeks and I had missed music greatly. I tried to believe that it was me who sat on the deck of his own ship at anchor, after a sea voyage of 1400 miles, and not the character of a book I had read. What is real? I thought a lot too about going on alone.

There had been times during the voyage when I had said that I would end it in Gibraltar, selling *Sheila* for whatever I could get, and take a passage back to New Zealand. No, there could be no turning back, no giving up, and I thought deeply how best to express my reaction. Before sleeping I wrote in my diary: 'I would not do that voyage again for five hundred pounds, but nor would I have missed it for a thousand.'

3

*Time in port—visitors—the Gibraltarians—
a Bull Fight—departure from Gibraltar*

It was heaven lying in bed that first morning and having no care about the wind, the sails, the navigation, the shipping or anything else. I unshipped the dinghy from where it was lashed up-turned covering the skylight, rowed over to a big Fairmile and asked the crew to keep an eye on *Sheila* for an hour or so while I went ashore.

Before leaving England I had gone to the head Met. Office in Harrow, and the Cruising Club Library in London, and studied world charts of ocean winds, the yearly seasons, ocean currents, hurricane tracks, distances, and ports with repair facilities, particularly those inside the sterling bloc. From this information I worked out the fastest voyage that could reasonably be expected to New Zealand under sail. Time only mattered because time was money.

Travelling from west to east is the wrong way to go round the world by sail, because the earth spins that way and so most winds and currents are adverse. Many told me it was not feasible, but as ships had gone everywhere under sail before the invention of engines I assumed it must be feasible; besides I particularly wanted to revisit India, and Malaya where my regiment was still stationed.

The one deadline was to reach Aden by mid-February to catch the fair-weather Monsoon (the north-east) across to southern India, so in my planning I allowed about four days in each port to replenish supplies, and a week where slipping was planned. A fairly full description of how time passed in Gibraltar will show, as it showed me, how impracticable this allowance was, and also that a single-hander needs the help of a companion more in port than at sea, where the ship looks after herself for a lot of the time.

First I went to the bank to draw some cash, then sent cables home, and collected a pile of letters—from home, friends, strangers asking for news of the voyage, Army correspondence concerning pension and resignation, *Sheila*'s insurance, literary people, official queries about my ration permit to draw bulk supplies before sailing, and the Income Tax people.

A very great friend of mine in London, whose letter I eagerly opened, merely wrote,

> ... only a few words to let you know I'll write when I get your cable. A long letter is pointless now when I doubt if you will ever arrive in Gibraltar to receive it.

That was from John, who had been Adjutant when I joined my Gurkha regiment. In those days we had held diametrically opposed views on how subalterns should behave.

There was no letter from my sailmaker, with whom I'd placed an order for a new mainsail together with its payment before sailing. He had told me that he could make the sail and get it to Gibraltar by the time I arrived. There was no sail and no letter explaining why not.

All those letters had to be answered and posted, and others written ahead along the route to confirm previous banking arrangements, and to request the re-direction of mail up to a certain date to Gibraltar.

Amidst this I was working on an article covering the voyage for an English yachting magazine and a Sunday paper, trying to get it away before any local agent for English papers came to interview me. He would cable the story back before the plane bearing mine had even taken off.

These newspaper interviews en route were a persistent rival in my own field because papers never pay for these interviews, always expect the full story, and then of course publish locally and at home too. The usual payment for such an article is about £10, which is hardly worth the time involved. However, I sold my first article locally for £2, and nowhere else, which indicated a very low standard of living for the future.

Then came attention to *Sheila*, because however careful I was at sea she always needed much attention in port. First a good scrub out from stem to stern below decks, including the bilges which always brought to light lost articles from toothbrushes, tin openers, to half-peeled potatoes. Masses of damp clothing and bedding had to be washed in fresh water (salty wet clothing never dries), dried, and re-stowed. There was always mending to be done because life at sea is surprisingly hard on clothing, but in most ports some wife ashore kindly undertook this task, to earn jealous comments from a neglected husband ('She'd never darn a hole like that in one of my socks!'). My shore-going clothes (a shark-skin dinner jacket was kept untarnished for the whole voyage) and spare linen had to be unpacked, aired, and repacked; laundry had to be listed, delivered, collected, and re-stowed; the galley with its pots and pans had to be brought up to visitor-standard and the engine rubbed over.

There was less to do above decks because this was more carefully watched at sea, but in port I checked every rope, block, shackle, and cleat alow and aloft, and changed them where any doubt arose. Long seams of the mainsail had to be re-sewn where chafe had weakened the stitching. Technicians were called in for those tasks beyond me through lack of tools or knowledge, usually to do with electricity about which I knew nothing and regarded as magic.

I nearly always lived on board in port, because apart from saving rent meals prepared by myself cost about a third of what they did ashore. However, daily marketing, cooking, and the cleaning-up afterwards took up a surprising amount of each day. I developed a genuine respect for the unsung housewife, and whenever I did think of enlisting a crew it was to relieve me of this work more than any other. Feminine tenders for the post were received at various times, but it would be impossible to confine a woman entirely to the sphere of cookery and washing-up; outside that, if she were dull it would be difficult on a small boat, and if attractive you'd do the washing up anyway.

The next job was to check those stores left, estimate the balance required to make up sufficient for the coming voyage, order, collect and stow. Charts and the Pilot Book for the next stage also had to be bought from the local Admiralty Chart Depot.

Another need somewhat neglected was rest. I had thought that a couple of nights' good sleep would repair the physical fatigue of the voyage completed, but something more than this was needed because the mental stress of this venture was far greater than the physical. I often found it very difficult to get myself started on the various tasks, not that I've ever been enthusiastic about cleaning bilges or checking laundry.

A further time-devouring factor in port which I could not avoid was the social aspect. Sometimes in the midst of work I'd see a person nearby staring at *Sheila*, obviously in a dream. He was sailing her into a calm lagoon, could already taste the first cool draught of coconut milk, and smiled at the laughter of dusky maidens clambering on board—so I'd ask him if he would like to look around. Others would introduce themselves and show genuine interest, others would be too shy or too aware of my preoccupation to do so. From all these I found many friends. The infuriating people were those who jumped on board uninvited, and said, 'Hello, quite a ship you have here. Can I look below?' and once ensconced in the cabin talked unceasingly about themselves, as if conferring on me the privilege of listening.

'I've always planned a trip like this, but I'd have a bigger boat and a cutter—they're easier to handle. Where are the spare sails? Oh, I'd have two spare mains anyway, you can't afford to take risks. And have petrol tanks built in behind the berths too. You came direct from England? I'd call in everywhere—I'm very interested in places, you see, not only my own home town. I like to get around. Magazines would lap up stories like this—and advertising—hell, you could pay for a trip like this if you worked it right and still come out on top. I know how to look after myself all right…' and so on until I was nearly screaming. Last would be conferred the greatest honour, 'Do you know for two pins I'd come with you.' Amazing.

One night at a small cocktail party of English people I said, 'I want to meet Gibraltarians. How do I go about it?'

'You're new here, old man, and may not understand. We have nothing in common and there's very little mixing. They're quite different.'

'Without any implication,' I said, 'that's exactly why I want to meet them.' There was quite a silence, but a Cable and Wireless member came to my aid later and this led to my happiest memories of Gibraltar.

Gibraltarians have all the charm and vivacity of Continentals. They took me to their beautiful homes high up the Rock face (there are several millionaires in Gibraltar, and the rest of the company were doctors, solicitors, or in the big import-export concerns which flourish there), and it was very pleasant to get reasonably well dressed, have a delicious meal in the atmosphere of a cultured household, and to relax in the warm evening talking of books, people, the difference between countries, the history of Spain, and listen to a superb radiogram.

Some Sundays they took me deep into southern Spain when we left the cars and walked through the great cork forests, over old roadways still showing signs of Roman construction, to view a massive squat village like a fort, built on a hill-top by the Moors and looking similar to Pathan villages on the North-west Frontier of India. After a picnic lunch by a stream and a siesta with a glass of cold beer, we drove to a beach near Algeciras and had a swim. In the evening we ate *al fresco* under the green canopy of great trees at a country inn, the *Meda Flores*.

Several evenings I went over to the Spanish town of La Linea to see more of Spain in my own time, and to eat Spanish food—which I failed to get, because all restaurants catered for the hundreds of troops and sailors who, it seemed, always ordered steak, eggs and chips.

I sat at a café with an excellent sherry at tuppence a glass, and watched the world taking the air. The streets were packed with people strolling up and down, leaving no possibility of any movement by motor traffic and nor did any attempt it. The youths looked over-dressed and mostly effeminate to our way of thinking, but the girls carried themselves beautifully, and the prominent ladies of our television world would have passed unnoticed among them.

One Sunday afternoon I went to see a bull-fight, partly to try and understand this much discussed sport and partly because I'd never seen one before. I was surprised to learn that there is a strong religious aspect behind it, dating from a time many years ago when a Pope ordered that it must cease, unless a good reason could be given for its continuation. The reply which gained this permission was that the fight between man and bull represented the fight between good and evil. As all things demonstrate this anyway, it still seems a bit hard on the bull.

There were the usual six bulls killed that afternoon, and to me the cruelty and odds against the bull were more apparent than the skill or courage displayed by the men. This is not a fair criticism; in the same way, anyone without a knowledge of boxing, seeing a first-class fight for the first time, would judge the brutality greater than the skill demanded or displayed. In any case, it is entirely the Spaniards' concern how they amuse themselves, and their resentment of foreign criticism is understandable.

The climax of the experience came for me as the last bull, weak from wounds and loss of blood, from the exertion, from the frustration of its efforts to grapple, stood with lowered head before the aimed sword of the matador. A quiet Spaniard beside me whispered, 'We call this the Moment of Truth.'

Sometimes a quiet casual remark made with no intent for effect strikes you like a douche of ice-cold water. The Moment of Truth—it so exactly described an experience during the War when I had known that at the end of four seconds I was to die. It is then that one is shown the Truth, with the shattering realisation that suddenly it is too late to do anything about it. The last chance has gone.

However, I survived but those years after the War had been restless years and by 1949, when I resigned my Commission in the Regular Army, I was sick to death of having my life tied to events over which I had no control, of divided loyalties. Leave back to England had not helped, just as the first post-War leave failed many others. Strangely, to live safe and clean, to be back after so long with those we loved, had brought little peace, and it was disappointing and bewildering to those who had welcomed our return. So often one heard the kindly whispered advice, 'Poor chap, just leave him alone. He had a terrible time during the War.'

I have often wondered, when people return after a great unhappiness, or illness, or period of trial and humiliation such as being a PoW, whether their restlessness, indecision, and strangeness is in fact due to nerves. It used to be called shell-shock and is now called battle fatigue. Perhaps it is because these experiences teach a man a revision of values he finds almost impossible to live in our normal lives. It takes time to forget what he has been shown, to dull the memories, and this is called 'a period of readjustment'. It certainly is. He feels a traitor until time deadens the memory and the logic of practical living re-asserts itself— if he lets it.

At this time more than any other a man needs love, a complete trust to whom he can bare his soul without fear of ridicule or exposure; without this he will try to hide from it, and the results may be disastrous both to himself and to those he loves most.

It does not need a psychiatrist to give the answer to this simple problem, and I determined on a way of life where I would be beholden to no one, with no conflicting loyalties forced upon me, with the maximum freedom to shape my own affairs. And so I resigned my Commission and deserted my Gurkha troops, men who had given me friendship and loyalty over years of war and peace. It is strange indeed how the attainment of something good can be the greatest menace to the attainment of something better.

4

*A strong hand—a shy shark—an escort of whales—
a lucky sparrow—arrive Algiers*

Just before noon, on 14 October, after nearly four weeks in Gibraltar, I said goodbye to many who had been so kind and, using *Sheila*'s engine out of the camber to pick up the wind outside, headed for Algiers. My friends drove out to the extremity of Europa Point, and as I sailed close inshore their white handkerchiefs waved farewell, and their faint voices carried *bon voyage*.

Algiers lay about 400 miles to the east, and I planned my course via the small isolated island of Alboran (100 miles east of Gibraltar and mid-way between Europe and Africa) to Cape Ferrat, where the African coast bulges northwards. After that I would continue along the coast about twenty miles offshore to Algiers.

My reasons for going to Algiers were, first, that I admire the French, secondly, that the distance was just right for one more stage of my journey, and lastly because several people in England had said, 'Don't go to Algiers. The French there are very anti-British and will be most unpleasant to you travelling solo.'

None of these informants could give me a reason for this antipathy, and so I decided to go to Algiers and find out for myself. Antipathy does not exist without a reason.

It was strange being alone at sea again, and as darkness fell on that first evening I felt very much on edge. Flukey winds demanded continual sail changes, and belts of rain reduced visibility to make my position on a main shipping lane dangerous. The second day brought all the warning signs of a coming Levanter, the easterly gales described in the Pilot Book, '...from October to November Levanters bring strong thundery disturbances much feared by sailing craft... it is at this time that water-spouts are most likely to form.'

The light on Alboran showed up dead ahead after dark on the second day, and for the next two days I lay hove to under storm canvas as the gale raged. Alboran itself is a low, flat slab of rock, which from a distance looks very like an aircraft carrier. The lighthouse is surmounted above a squat double-storeyed stone barracks which is the centre of the Spanish convict settlement, and the only other signs of habitation visible were two football posts, standing forlornly in the arid isolation above the waste of sea.

At dusk two days later I was drifting too close to the Spanish coast, and tired of waiting for the storm to end I decided to sail south to Africa in search of better weather. The wind was hard and steady on the port beam and I let *Sheila* sail herself, the helm free, because in rough weather she has an uncanny knack of slipping round the crests in a way my own helmsmanship has never mastered. It was a wild, beautiful night and I sat in the shelter of the hatch with a mug of coffee, for once feeling more like the owner than the paid hand. By 2 a.m. we had long since crossed the shipping lane and I went below and slept.

A strong hand grasped my shoulder, shaking me with firm decision, and a voice said, 'Get up, there's a ship coming.'

I obeyed immediately and there down-wind was a black mass, port and starboard lights aglow and the two mast-head lights almost in line, indicating that our courses would cross. She appeared to be about half a mile away, which at fifteen knots is only two minutes in time.

My starboard light faced her, but the absence of a mast-head light should have told her that I was a sailing ship and so had the right of way over steam. To make sure, I played the powerful beam of the Aldis over the sails, throwing them into vivid relief against the darkness, and then signalled the bridge of the coming vessel. She held her course, and only when the flared bows towered over me did I force the helm hard up; her bows plunged into the head sea, the wave nearly swamped me, the canvas flapped wildly, fit to disintegrate, and as I slid past only yards from the tall iron side I saw a head peering down at me from the end of the bridge. With all my strength I yelled upwards, 'You rotten bastard,' but with the noise of turbulent water my cry was lost in the wind.

By dawn I was off Cape Tres Forcas on the Spanish Moroccan coast, and for the next three days lay becalmed with only an occasional breeze. After the buffeting and wetness of the preceding week these days of rest and sunshine were paradise. I usually had a picnic lunch in the cockpit after a swim over the side—it was a thrill to stand on the end of the bowsprit and gaze fathoms deep into the clear green water, and then plunge like a spear far down into the sun-lit depths. I had no worry about sharks. I had always been told that there are none in the Mediterranean.

During the evening of that first day five small fishes joined me, about eight inches long, coloured an electric blue with dark maroon bands encircling their bodies. I caught one next morning on a small hook and had it for breakfast, finding it tasted very like trout. The others stayed with me, swimming in the bow-wave as *Sheila* sailed herself at about three knots to the north-east, and darting out occasionally to inspect some marine tit-bit. I wondered what these fishes were.

The wind had almost died by eleven o'clock, but I noticed a distinct eddy about thirty yards astern, and was puzzled that the wake should extend so far when the ship moved so slowly. However, it was a beautiful morning, so I stripped off and climbed out to the end of the bowsprit, but paused because of a very definite feeling of apprehension. There is a reason for everything, so I went back to the cockpit to consider it. Although *Sheila* was now barely moving, the eddy was still there and I knew it was this that worried me; it didn't seem natural. I threw a bucket of drenching water over myself instead of having a swim, feeling rather old-womanish for being 'windy'.

During lunch a glance across the cockpit showed the eddies just under the quarter, so I looked over the side—to see an eight foot shark, and fussing round his nose were small fish like the one I had eaten for breakfast. Pilot fish, of course. That shark had probably been there astern since the pilot fish had first joined me the previous evening.

Those nights off Tres Forcas were so beautiful I usually stayed up late, sitting on the foredeck with my back leaning against the mast. There was a bright moon, and when a breeze came the sails slept in graceful curves above me against the night sky. There was a sense of great peace, a quiet happiness, and I wanted to stay awake to be grateful for it.

About two o'clock in the morning I awoke to a loud explosive puffing noise, like heavily-laden trains pulling out of a station. Six small whales about *Sheila*'s length lay on the surface right alongside, blowing lazily. Soon others joined them, and as a breeze came they eased in around the bow, flank to flank and moving ahead with the ship as if talking to her. I crept carefully forward and lay full-length on the foredeck, my head over the bow. The flank of the whale beside me was so close I could have stretched out and touched it. I counted twenty-one.

The escort stayed for about half an hour, their black, wet backs glistening in the moonlight and the breeze carrying the dampness of their breath over me. Others were about a fathom immediately below, their blunt heads and broad shoulders etched in phosphorescence, until moving forward they nudged their fellows aside to rise and blow. Some beside me occasionally humped their backs to submerge the head, and wave the long slim tail and graceful flukes high in the moonlight.

They brought no sense of menace. It was as if they had some family affair to discuss with *Sheila*; I was excluded but my presence tolerated on her assurance of my good behaviour.

Next morning a school of porpoises joined *Sheila*, and were either playing some game which allowed the greatest intimacy and demanded the highest standards of aquatic skill, or else were very much in love. They circled the bow in pairs about six feet under water; one nuzzled the other which, after playful and unmistakably feminine protest, turned over on its back to show the startlingly white belly. The other moved over her and together they swam in graceful curves and circles as one united being. Still together they rose upwards, and only just before breaking surface the female turned upright and flank to flank in perfect harmony they emerged, blew, and dived as one. As they entered the water the female again turned gracefully over as she curved away, and the male covered her again. The grace and perfect harmony was beautiful to watch.

At noon wind came, and over the next twenty-four hours *Sheila* did a nautical striptease as it steadily increased to gale force. I took in the mizzen in mid-afternoon as we passed Alboran, and turned slightly north of east to head outside Cape Ferrat. At dusk I changed to the No. 2 jib (a precautionary measure), put one reef in the mainsail, and ate a large meal from a pot of stew prepared that morning on the first sign of coming wind. This meant that for the next two days good meals were ready prepared and only required heating.

At 10 p.m. the wind was still increasing so I put in the second reef, and at midnight changed to the storm jib, put the third reef in the main, and had a large mug of hot sweet coffee. At three in the morning the sail area itself was safe enough, but I was running too fast in the big following seas, as the ship told me by the pressures and vibrations on the helm, so in came the main altogether and up went the mizzen. With the first pallor of dawn I took in the mizzen, and the force of the gale in the rigging and bare poles continued to drive *Sheila* before it at about five knots.

After a hurried breakfast, the hard, dry feeling which sometimes comes to my eyes after long periods without sleep made me realise I could keep going for ever. Everyone knows its opposite, such as after a good Sunday lunch, when a worthy and authoritative relative insists on doing the crossword, and only the greatest effort of will can keep your eyes open.

Soon after this a wave lifted *Sheila*'s stern in a way that made me take a hurried glance over my shoulder—it was a giant as were many others, but also a huge concave curve of lucid green water, the sun shining right into it. Usually only the crest of a wave is curved, but in this the whole body from the base to the top was one huge cave, only seconds away.

The stern continued to rise as we forged ahead with the speed of the wave, until I was staring vertically down at the bow. The masts seemed to be almost horizontal and I could feel *Sheila*'s struggle not to nose-dive. For one second of time it seemed that the stern was going right over, somersaulting the whole ship, but the flare in the bows had just enough lift to hold her until, with the stern high in the air, the crest passed beneath in a roar of broken foam and *Sheila* sank back to her normal position.

In all the voyage that one wave was unique, and never again did I meet another like it. Unfortunately for my future peace of mind I did not know that at the time.

The wind veered towards the north later in the day and eased, becoming wet and gusty. Far ahead in the grey mistiness I saw a tiny black speck, and after a few minutes came up with it: a lone sparrow struggling against the wind, heading back towards Europe. This was indeed flying in the face of Providence.

Off Tres Forcas many sparrows had passed me, flying from the bitter cold of Europe to the warmth of Africa, which meant that many must have covered about 200 miles of open sea. I supposed that this one had come so far, given up hope of finding the warm land his instinct had said existed, and had turned back to recover the land he knew.

'Hi, you're going the wrong way,' I yelled, 'come and rest awhile.' (I always talked aloud to birds and other living creatures at sea; they may not have understood my words but I'm sure they often understood the meaning.)

For a while the sparrow continued the struggle, keeping just ahead of *Sheila* but making no way to the North; then suddenly he turned back and landed on the deck beside me, chirping loudly. I knew exactly what he was saying—he was swearing like a trooper.

'Now listen,' I said. 'I probably know more of ocean crossings than you do, and it's no use making a big decision like turning back when you're tired. The coast is bulging north all the time and tomorrow morning I can drop you off in sight of land. You're welcome to stay the night.'

However he took off against my advice, and this time made enough headway to disappear in the darkening mist. I felt quite sad about this because I knew he wouldn't make it, and it had seemed such a stroke of luck that *Sheila* had arrived in that one bit of ocean in time to help him.

But half an hour later he was back again, very upset, and not quite so undaunted. I was sitting on the cockpit coaming with one foot braced on the other side to give me more control of the helm, and the sparrow hopped on to the seat under my leg, sheltered from the rain and spray. He was still cursing hard.

'Hell,' he fumed. 'I left Europe on a silly hunch, have flown for miles and miles, admitted my stupidity and turned back—but do you think this lousy wind will let me get home again?'

After a time he cooled down and went on a tour of inspection. He flew on to the hatch, nearly got blown off, so hopped under the upturned dinghy; then he flew through the hatch into the galley, and on into the saloon where he perched on my books and got very drowsy. As darkness came he went back to his perch on an inverted seat inside the dinghy, but every now and then popped out to cock an eye at the weather, obviously thought it horrible, and went back to his snug perch.

I hove to for supper, and put some water in the lid of a cigarette tin and some biscuit crumbs under the dinghy for this small guest who had apparently decided to stay the night.

I'd not been able to get sights since leaving Alboran three days before, and being very tired I began to doubt my navigation. I knew roughly the distance covered but I was afraid of passing Algiers altogether, so that night moved in towards the coast to pick up a light. Besides, I wanted to ensure being in sight of land by dawn so that the sparrow would see his way. Tenez, about eighty miles short of Algiers, showed up about midnight, and putting *Sheila* on the starboard tack to drift offshore I slept in hour spells, poking my head through the hatch to check between whiles.

I sailed again from about 4 a.m., and with the dawn and the watery sun my guest hopped out on to the hatch to inspect the day. The coast was only a few miles off the starboard beam.

'There it is,' I said, 'straight down-wind; you can't miss it.' The sparrow gave a couple of perky cheeps and off he went.

It is strange to think of these little land birds following an instinct which tears them from their homes and drives them over 200 miles of open sea; others are lucky, of course, the ones who presumably fly down

Italy and take off from Sicily; but Nature arranges things beautifully for them all. At that time of the year the warmer air over Africa rises, and so the cold air of Europe is drawn towards Africa to replace that which has risen—in other words winds blow from Europe to Africa and help the sparrows on their way, so I suppose most get there safely. But mine had lost faith in that strange urge, and it was lucky indeed that *Sheila* happened to be passing.

The wind increased again bringing huge seas, and during the next night terrific gusts. How true the sailor's maxim:

> Rain before wind,
> Top halyards mind.

As the first fat drops fell on the deck I rushed forward and let go the halyards just before the wind struck. Even under bare poles the force of those gusts heeled *Sheila* far over; they flattened the sea and raised a thick white cloud for ten feet above the surface, the tops torn from the waves. Later information confirmed the gale and reported gusts up to seventy miles an hour.

I rounded Cape Caxine in the early hours of the morning and hove to off Algiers in comparative shelter. At dawn it was calm.

Algiers was far larger and more magnificent than I had expected. It lay in the shelter of a curved bay, great blocks of flats, hotels, business offices and cathedrals rising tier by tier up the hillside overlooking the built-up harbour. The sprawling mass of the kasbah (native quarter) lay at the northern end, gaining dignity from a magnificent mosque standing in its centre.

Owing to what I had been told in England of French hostility I was meticulous in my preparations for entry. Official papers were all ready, and with the flags for Health and Pilot (both of which the Book told me were compulsory) flying, I headed towards the entrance of Vieux Port to collect these officials.

Over the next four hours both passed me on their way to other vessels, ignoring my signals, until I was able to call a pilot who passed close.

He said a great deal very quickly which I reduced to, 'I'll be back soon,' but this was perhaps not so.

My rage was mounting and fully confirmed the advice received in England. I managed to hail a Customs launch, and after receiving clearance I asked for a tow back to harbour as the engine was out of action. The official asked, 'Isn't this a sailing boat?'

In those days I was very chary of entering a strange and perhaps crowded harbour without the engine at least ready for emergency. Winds are often gusty under such a sheltered hillside, leaving you bereft of power one minute and driving the gun'ale under the next, which makes picking up a mooring or going alongside extremely difficult single-handed in the confined space available in most yacht harbours. Besides, once people know that you are a lone ocean sailor, they expect you to handle your craft as if you controlled the winds to your own needs. And at that time I was tired, which always makes difficulties seem greater than they are; but that official's remark made me mad.

'All right,' I muttered, 'to hell with your regulations. I'll take her in under full sail, and it won't be the first time an English ship has entered one of your ports without *pratique*.'

Unfortunately the wind dropped just as I entered the breakwaters, and a current was carrying *Sheila* into trouble. I hastily unshipped the dinghy and began the back-breaking task of towing her. Several launches passed taking no notice until the Health Launch itself offered me a tow, which I declined muttering what he could do with the whole nation.

Darkness had fallen and I was getting very tired when two fishermen circled and took me in tow, and cast me off in the crowded yacht harbour. I was grateful for that. I put a light in *Sheila*'s rigging and let her drift in the harbour, while I scouted round in the dinghy for a mooring or a place to tie up, and then rowed back, took her in tow once more and made all fast by 10 p.m.

On the pier near the yacht club was an elderly man in old clothes, with a fine face seen as he lit his pipe. I couldn't decide whether he was the night watchman or a retired admiral, but he took me to Madame who was just closing up the club.

I wanted cigarettes, having left Gibraltar very short in order to cut them down. The scheme had worked, as I'd smoked none at all for the last week.

Madame was a handsome woman with thick black hair and very black eyes (I learnt later she was Spanish). I knew I looked pretty haggard, and immediately felt her response of sympathy and willingness to help in any way she could; my anger and resentment faded to leave me feeling rather foolish.

5

Arab independence—a cabaret party—uneasy departure

I awoke late the next morning and lay in my bunk as the sun streamed through the skylight, feeling very much at peace, rested and unhurried. I got up to make a cup of coffee and took it back to my bunk to enjoy with a cigarette.

I went ashore about ten o'clock in search of the club Secretary to ask permission to lie at the club jetty, which was of course private property. Two members in the club introduced themselves, said they were glad to have me there and to regard the club as *chez vous*. Later one of them drove me into the town to meet the British Consul, and to collect money and mail from the bank.

There was not the slightest suspicion of the antipathy I'd been expecting, only friendliness and offers of help. By listening to idle chatter about Algiers I had jumped to conclusions the previous evening, and might easily in anger have created discord where in fact none had existed. Later I asked about the entrance difficulties.

'But no one expects a yacht to be bound by such regulations.' The Port Authorities allowed yachtsmen to come and go as they pleased, relying on them not to abuse the privilege, which is a very reasonable attitude indeed.

The next few days passed in glorious idleness before tackling much the same programme as had applied in Gibraltar, but it was heaven to rise late, dress in clean shore-going clothes, wander into the town and sit at a cafe with a glass of lager, read an English newspaper, watch people pass, and return to the club for lunch. I had nearly all my meals ashore in Algiers, partly because the French often invited me as their guest, and partly because the discrepancy between my own and French cooking was too startling to ignore.

The overall picture of town and country gave a very interesting impression of what France has achieved in just over a century of occupation. When I was there, in 1950, Algeria was a rich country enjoying the advantages of a stable well-administered community, a state due to French brains and industry, and one which it is very doubtful that the Arabs alone would ever have attained. But even at that time the demand for freedom was heard, part of the great wave for independence sweeping over Asia and into Africa.

It is indisputable that the Arabs should have their freedom, but their freedom to what? Freedom to take all that the French have constructed and organised, to revert to a lower standard of living, to inter-tribal wars, to unjust oppression by their own autocratic rulers, or perhaps to find their own destiny in a land which they took from others before the French took it from them?

I remarked to some Frenchmen that the situation was similar to that which we had experienced in India.

'Not at all,' they said, 'India was never your home. France is not our country, this is our country. We were born here, our parents were born here, we've built our homes and made our careers here, and some of us have never been to France. This is our country by right of birth, just as much as it is the Arabs'.

'And has the Arab got equality of opportunity?' I asked.

'He doesn't want equality of opportunity, he only wants what he sees we have created by industry and hard work.'

For the next four years I was to pass through and live in countries where this problem was foremost in men's minds, and to see both the construction and disintegration to which it may lead.

The life at sea kept me very fit and after a good night's sleep ashore I longed for feminine company, for the sheer pleasure of taking out an attractive, well-dressed, intelligent, amusing woman, such as I danced with one night at a private party in a very nice flat. I could not ask any of those girls to meet me again because for all I knew they were 'going steady' with one of my hosts, and nor had I the money to entertain them as they deserved.

In some ways women are far more intelligent and stimulating to talk to than men. Men were usually most interested in the planning and practical hazards of my voyage, but almost without exception it was a woman who asked, 'Why are you doing it?' which is surely the most intelligent question of all. This question was almost an obsession with me at the time because I hardly knew the answer myself; but discussion with intelligent and sensitive women has been one of my greatest aids in deciding it.

Sheila had to be slipped in Algiers, scrubbed off and anti-fouled against marine growth below the waterline. Members of the Yacht Club arranged all this with the *Sport Nautique* across the harbour, right beside the *Café Portuguese*. The whole thing, including work and paint, cost me £5 (as opposed to the £50 quoted in the dockyard in Gibraltar), and I am sure that various members contributed to the cost which must have been nearer £15.

The first afternoon on the slips a French friend came to explain the need for me to sleep on board that night, but I was slow to understand. He tried English.

'Bad man come *dans la nuit, n'est-ce pas?* 'E pick-pocket *la belle Sheila,*' which made his meaning quite plain!

On return to my berth at the club I found a big motor-sailer beside it, her decks and rigging festooned with unmistakably feminine laundry. As I made fast, one girl after another peeked through a porthole, the hatchway, the cabin glass, to see what the disturbance was about. The boat seemed to be full of them, and all very pretty. Later the skipper called on me, a tall, dark, very smooth Italian who spoke English well. He had heard of my voyage and was interested in my ship, and I hoped that he would invite me over because I was interested in his also.

We entered the big saloon, and on his call the chattering in the forward cabins ceased and in walked five lovely girls. The German was tall and icy-blonde, the French petite and also blonde, one Spanish girl was dark and very like Ava Gardner in some of her photographs, the other Spaniard was a red-head with a gorgeous figure; last came the Italian girl, and she was beautiful. The skipper explained that they were a cabaret party who toured the length and breadth of the Mediterranean, and had just signed with the local casino. Would I like to come out with them that evening?

I was in the club later that evening when the cabaret party came in for their dinner, and I went over to ask them what time they were going out to the casino. On return to my party I met a concealed but very noticeable atmosphere of disapproval.

If such a yacht had tied up in England off a club, various local people would have offered any help, but the party would not have been made members of the club. The French did exactly the opposite, making the party members and so entitled to the facilities of the club, but thereafter they kept entirely separate socially, even in the club itself. They expected me to do the same.

This posed a very difficult problem. I did not want to offend people who had been very kind to me and taken me to their homes, but how could I meet other people tied up alongside and ignore them inside the club when with my hosts?

Paul took me aside (he was one of those people whom everyone liked for his sincere good nature and generosity) and said, "E is 'ow you say? A leetle man.' I tried to explain that I understood, but to me he was interesting.

The social circle in a good regiment is very carefully prescribed, and I'd never met people like these; it was half the reason for my trip, to meet as many people as possible, partly to discover what foundation (if any) colour, racial, caste, and religious prejudice had. I very much doubted whether that big yacht could be maintained, and that party clothed and fed, on the takings of a mediocre cabaret turn at out-of-season resorts; the rest may have come from smuggling, dope, or even white slavery. These were types new to me, and anyway the idea of dancing at the casino was attractive—is there any harm in that? So long as I didn't involve the French, I could not see that it had anything to do with anyone but myself. But they did not forgive me, as I learnt when I got to Bône.

I went several nights to the casino, watched the cabaret, played at the tables and won enough to pay for champagne, danced with the girls, and got back about four in the morning. One of the girls told me that she came from a small village and her people were very poor; then she met this party (I could not gather how), and now she travelled, saw other people and places, and had nice clothes. 'Is it worth it?' I asked.

She looked at me steadily and said simply, 'I lose nothing.'

It was the Italian girl I watched most carefully (we had no common language) as she danced with me, very noncommittal and afraid, as she swayed to a false hula half-naked on the stage, with fat lecherous locals eating her alive, as she sat at our tables between dances. She had a look of extreme unhappiness in her eyes as if her very spirit was in danger, and I'd only once seen that look in a woman's eyes before.

That had been a tall attractive English girl travelling from England when I was returning to India after leave before the War. She dressed beautifully, politely rejected any offers of friendship even from women, and looked—unhappy does not describe it; it was an inner unhappiness, utter hopelessness.

At Port Said, long after others had gone ashore and the deck by the gangway was almost empty, a short rotund Egyptian in a red fez and a loose blubbery face came on board, took her by the arm, and led her away without a word from either of them. It was one of those occasions when you feel something terrible is happening before your very eyes, and you ought to do something quickly before it is too late—and suddenly it is too late. It is sometimes the things you don't do that you regret even more than those you do.

The last two days in Algiers were the usual rush of departure and the good Paul stood by me to the last, providing transport which is always a need for last minute shopping, the bank to see, officials from whom to collect clearance papers, and farewells to say. I was unhappy leaving, knowing I'd hurt people's feelings but not agreeing that they should have been hurt.

On that last morning I nearly went to the local convent to ask their advice and help concerning the Italian girl, but what could I say when it came to it? That a girl was unhappy and I felt sorry for her? I did nothing and sailed that morning, feeling I was leaving something undone that I ought to have done, and the feeling has never left me.

6

Thoughts in the wind—arrive Bône

On the second night out I wrote in my diary:

> 9 p.m. It is a beautiful night outside, half-moon, light wind, gentle sea. I've found a station on the radio giving the most heavenly music; *Sheila* is sailing herself north-east, allowing me to be on my bunk listening to this music—it makes me feel good inside as if there is a great energy there, exuberant yet peaceful if that is possible. I'm very happy.

Having had no sleep the night before while close to land and shipping I slept that night at 11 p.m., leaving *Sheila* to sail herself still to the north-east at an angle off-shore. I awoke after an hour (which was by this time automatic) to find that the wind had gone from east to west, turning *Sheila* with it and sailing her back towards Algiers. It was a good wind so I turned back on course and stayed at the helm through that night until the dawn.

A solitude without loneliness was nearly always with me at sea, particularly over long hours of darkness when the ship and I and the waves and the wind all seemed to be part of the same thing. Thoughts then came freely and dispassionately, unmoulded by convention or prejudice of any kind, or even of upbringing, thoughts which on land within an atmosphere more of man than of nature would either not have come at all, or would have been driven away without consideration. In this state the hours passed unnoticed, steering became automatic, the mind did not register fatigue or not-fatigue, cold or not-cold, and like a vacuum it was empty to receive thoughts which sometimes came of their own accord and were sometimes induced by myself. How do you *know*?

This struggle to know had sometimes become a searing pain, and brought periods of restlessness and frustration, to escape from which had always embroiled me in trouble of one kind or another; I would hurt someone deeply, or simply get drunk and 'put up a black', as we termed such digressions in the Army. But sometimes came an answer, an understanding which brought a deep peace, a sheer heaven.

'Have faith,' had said a padre to whom I had spoken of the need to escape that restlessness, that hell. But I had no faith, I who asked why, because to ask why is to doubt and one iota of doubt destroys all faith; and who was I to have faith when even Peter who knew Christ personally had not the faith to walk across a few yards of water to meet Him?

'Believe,' they say, but you cannot—or I cannot—believe a thing just because you want to; and so can I derive no comfort from the orthodox insistence that I do so.

Thus the hours slipped by unnoticed. Ships with their blazing lights passed unaware, stars I saw nightly rose in their allotted paths, climbed to their zeniths, and sank astern. *Sheila*'s tall mast swayed as always before me, the sails curved upwards in the starlight sleeping with the power they held to drive us east, and the waves came endlessly with their quiet murmur at the bow. And so I stayed all night at the helm with my thoughts in the wind.

I passed the Gulf of Bougie and arrived off Cap Bougaroni at dawn five days out of Algiers, and over half-way to Bône. My diary records:

Off Cap Bougaroni, heavy gusts, rain, big confused seas; terribly tired and more listless than I ought to be, my arms and back aching from steering in the big following seas. Could have used the full main all yesterday and last night except for these fantastic gusts, when even the trysail is too much unless I run before them to reduce the pressure. Bloody cold too.

Headed straight across the Gulf of Stora—45 miles—and arrived off Cap de Fer in the afternoon. It's a wild coast of deep hidden inlets and craggy headlands, easy to tell myself stories of the Barbary pirates and a thrill to sail over the same waters—imagining a long, slim, dark vessel streaking out from behind a headland to pounce on a cumbersome merchantman, the fight on the decks, breaking open the loot, the pick of the women in the captain's cabin and all hell in the fo'c'sle! Life certainly had blood in it then—our nearest approach now, I suppose, is pinching a box of matches from the lounge and petting on the boat-deck of a P&O.

Hove to off Cap de Fer and cooked up a big meal of sausages, tomatoes, boiled cauliflower and potatoes. I didn't want food but was so listless—and frightened too—that I knew the last two days of semi-starvation were beginning to tell. I also cleaned up the ship and the sink-full of dirty dishes—difficult with *Sheila* standing on her ear half the time, but the old, old lesson—dirt and unwashed dishes depress one more than bad weather; after a meal and clean-up the waves were half the size and the wind half as strong.

Sailed down the coast all night and arrived off Cap de Garde (Gulf of Bône) at 2 a.m. 23rd.

My only reason for going into Bône was because the very kind President of the Yacht Club in Algiers had said (before the cabaret incident): 'You must go into Bône. I have a very great friend there who would love to meet you, and you too would like meeting him very much.'

Fair enough, but I had no charts of Bône harbour; those who knew it said it was simple ('Straight on, first left, second right, you can't miss it!'); there were the usual entrance lights which could be picked up while well out at sea, if approaching at night. The entrance to a harbour looks very easy to one who has seen it in daylight, but arriving out of the dark, even detailed instructions leave it difficult for a stranger. A lesson I learnt was always to collect all information possible, but the final responsibility for your ship is your own, so never economise on charts. I eased round Cap de Garde keeping well offshore, and then sailed in towards the city lights; if I hove to till dawn in that area I would have to stay awake, which seemed silly with a safe harbour half an hour away.

It is very difficult to gauge the distance of lights when approaching from darkness, and when I heard the roar of the heavy surge on the breakwater I moved in closer on soundings with the lead and line until I picked up a pair of red and green flashing lights, the usual entrance lights to any established harbour.

I prepared *Sheila* in the usual way for entry, took in sail because the wind under the land was even more gusty than outside and I didn't want to get tangled up in a steamer's mooring lines in the darkened harbour, plugged in the Aldis lamp, started the engine, and headed for the entrance lights. 'All that worry,' I said to myself, 'and it's as simple as this.'

A lesson learnt in the War applies as do many others to handling a boat at sea—when very tired be extra careful. And that night, although the entrance was clear ahead, I ran the engine very slowly, put it into neutral every thirty yards, and went forward to the bow with the Aldis to ensure that all was clear (from the cockpit the beam only reflects back from the rigging, making anything ahead impossible to see). I had no doubt about the way but sometimes local people set fishing nets in an entrance channel, or anchor small craft there, and it makes a bad impression if you enter a foreign port over the wreckage of several prized sailing dinghies.

It was pitch dark except for the flashing lights and the yellow glare of street lights beyond, and *Sheila* was 'smelling the bottom', telling me by the change of her movement to the seas. On my fourth trip forward with the Aldis, to my horror the beam showed a long breakwater of huge concrete blocks running straight across the bow, only yards away. I rushed aft and slammed *Sheila* into full reverse.

The red and green flashing lights were those showing the passage from one inner harbour to the other; being at an angle and higher than the unlighted outside breakwater, they looked from seaward as if they were the entrance through it. I later pointed out the danger of this to a French official.

'Not to anyone with the right charts,' he said, against which of course there is no argument.

By following along the breakwater I found the correct entrance into the Avant-Port. The surroundings were in darkness, but I picked up a steamer's mooring buoy in the middle of the harbour. It is against the law to tie up to one, but I'd had enough excitement for that night and only wanted to sleep; so I made fast and did so.

7

*'Monsieur, dormez-vous?'—'Have a drink?'—
Le Café du Bateau Plaisir—departure from Bône*

I awoke still tired and my first thought was to move *Sheila* from the buoy before some worthy official drove me from it; my second thought was to stay in bed and sleep until he did. That lasted exactly five minutes.

That day I completed all essential tasks such as seeing officials, cleaning out the ship, and drying clothes. I moved to a berth alongside the Yacht Club, and after an early supper went to bed and was asleep by eight o'clock. At eleven o'clock a feminine 'Ooh-hooo' awoke me. It was repeated several times, but knowing it was not for me I turned over to seek sleep.

Sheila rocked gently as someone came on board, and a small curly head was outlined in the hatch-way against the sky. '*Monsieur, dormez-vous?*' It was the most feminine '*Monsieur*' I've ever heard—it held concern for waking me, laughter because it was rather unorthodox to do so, anticipation, and doubt. I had no idea whether the visit was official, friendly, or commercial.

'Mais non, Mademoiselle.'

So Mademoiselle explained at great length with typical French vivacity, leaving me charmed but still mystified. Then a man came on board and explained that they had read of *'le navigateur tout seul'*, and would I come ashore to have a drink with them—two girls and himself?

The things I do for England—I'd only had about five hours' sleep in the last three days, but with international concord in the balance I had no choice.

It has often surprised me during this voyage how much more quickly mutual friendship is attained with people whose language is barely understood than with strangers of your own race. Perhaps because with foreign words you search more deeply for the meaning that lies beneath them, until minds fuse in an understanding when words become unnecessary, whereas in your own language you thoughtlessly assume that each of you mean the same thing by the use of particular words. It is common to have a lengthy argument on some subject, only to discover that your disagreement was not on a point of view, but on a different conception of the meaning of a key word. For instance, the word 'freedom', so freely used by political leaders and underlined in newspaper headlines; so many people have said to me: 'What a glorious time you must have in port, complete freedom to do as you like,' which shows that the meaning of this common word is not understood. Freedom demands far more control and sacrifice than any combination of man-made laws, parental discipline, or social convention. And again, 'Do tell me, have you ever loved during your voyage?' can mean anything from a crumpled pound note on a sordid dressing-table to the highest attainment of which we humans are capable.

In the morning I rang up the person I'd come to see, my reason for coming to Bône. He was very abrupt.

'Do you want any help?'

Not in that tone of voice I didn't.

'Well, if you do let me know,' and he hung up.

Obviously the French in Algiers had sent a second letter revising their first opinion of me, and this upset me very much. It is surely sad when a friendship developed on mutual esteem and respect is cut short by death, but how much sadder when it is cut short by misunderstanding. I may have been unconventional, which can be sometimes unimportant and often constructive, but I had not behaved badly. Their opinion of me is not important perhaps, but it is important that they should not feel their kindness to a stranger was casually abused.

The main street of Bône is really two streets with a wide avenue of trees between them. In the evenings I usually walked the mile from the port, bought an English paper and sat under the trees at a table with an aperitif.

One evening suddenly an American voice said, 'Hello there. You're reading an English paper so I guess you are English. On your own? Have a drink?'

We sat and talked for some time, or rather I listened while he embarked into the story of his life—he lived in Naples, had married an Italian girl of surpassing perfection as a woman, wife and mother. He also told me of his wonderful job in which he circled the Mediterranean ports—'...and it's a great set-up. In each place I now have a woman with her own lodgings who puts me up, sleeps with me, does my washing, and it costs me not a cent, but I get my travelling allowance just the same.'

I couldn't think of anything sensible to say, so I said nothing.

'Now don't get me wrong,' he said, 'I'm very fond of my family, and want my kids to have everything I didn't. Why, when I leave here I'll take a whole car-load of toys back—what do you think of that, a whole car-load?'

'Won't you spoil them?'

'Aw, you get me wrong again. I don't spoil 'em. I only got to speak real sharp to my young son of six and he's so scared he wets his pants.'

His views on the negro problem in America were quite definite.

'There's no problem—down the South where I come from they know their place and we don't let 'em forget it. They get off the street when I pass, I tell you.'

'But nowadays,' I asked, 'aren't many of them top-notch musicians, university educated, lawyers, doctors, and so on? In fact, more stable and constructive than many white Americans?'

'Aw don't give me that. They're black, aren't they?'

It is no use arguing with a person like that and one opinion does not represent all America's; but I had seen letters in *Life* printed after the Detroit riots expressing exactly the same sentiments, and admitting no reason whatsoever for holding them. Considering this attitude how can anyone blame Paul Robeson for turning to communism? Not because communism is better, but because as he saw Democracy it was barbaric. I understand also that his records are no longer sold in America, and to deny the beauty of a man's voice because his political opinions don't coincide with your own is as immature as refusing to recognise the established government of a people because it is not the government you yourself would choose.

After supper I always walked back through the dock area of warehouses, tall stacks of timber, and the lanes between huge bales of some bulky merchandise. There were very few lights and I always keep my eyes and ears open in such places, but I was told this area was dangerous—lone seamen on the way back to their ships were always being beaten-up, robbed and sometimes murdered by Arab toughs.

'Take a taxi,' I was told, 'with a French driver; with an Arab driver you never know where you'll end up.'

So the night before sailing I walked over to a taxi, the driver of which happened to be an Arab, and asked to be taken to the Yacht Club. (To be truthful, as most Frenchmen are swarthy and Arab taxi-drivers wore European clothes, I often found difficulty in knowing one from the other.)

He did not understand me and I had forgotten the French for a yacht club (*Cercle Voiture* or *Nautique*), until I finally tried, '*Savez-vous bateau plaisir?*'

'*Bateau Plaisir? Aaaah, oui, Monsieur.*' He slammed the door and off we went.

Half-way to the Yacht Club he suddenly swung left, then right, and we were in the kasbah. Up to that time I hadn't been in the kasbah (one native quarter looks very much like another and I'd seen plenty in India), so I let this entry pass, wondering where it would end.

The streets were unlighted except at odd intersections, the tall buildings crowded in and narrow alley-ways shot off to right and left in dark tunnels through the walls. Some of these old Arab buildings are many storeys high, and being built without concrete and reinforcing, the walls are very thick at the bottom, so that all doorways and windows are deeply inset. The street became so narrow that whenever we met a pedestrian between doorways, he had to climb into a window-niche to let the taxi pass.

Soon the taxi stopped at a deep-set doorway with a dusty light hanging over it. In faded begrimed letters under the light were *Café du Bateau Plaisir*!

The taxi-driver knocked on the heavy door; soon a grille opened and a heavily painted face questioned him as to my reputation. On his recommendation bolts were drawn and we were allowed inside.

Madame herself gave an exact impression of a *bateau plaisir*—one of those broad pleasure craft untouched by the careless crowds she carried, by the sweets, the discarded papers and the spilt beer, but all flags flying—she wore a dress with frills hanging from all over it.

We went through a hall and down some steps into a large open room, obviously for dancing; around the walls were a few tables and chairs, in one corner a red and chromium bar, and beside it a narrow stairway on which there was more movement than on the dance floor. My fellow patrons were a mixed bunch—foreign sailors in singlets and blue dungarees, Arab youths of the spiv type, a few Italians and Greeks, and several older negroid types in semi-European dress, drunk and excited by the availability of white women.

The girls were all white, probably Hungarian, Roumanian, Greek and French. They were variously arrayed in sweater and short skirts, or shoddy silk evening dresses with nothing underneath. One had a mere shawl draped around her, and others were in BBs and panties. It looked a bit rough, and so I invited the taxi-driver to join me in a drink at the bar as a first line of defence. He was already under attack when a hold on my arm turned me to gaze down into a heavily painted face, blue eyes, long nose, and thin lips—I could see the lipstick smeared over the skin to make them look fuller. The stale paste around the corner of the eyes and lips was cracked. She leant closer with a gesture of affection, almost as if realising her need for just that, bringing a whiff of stale scent, brittle dead hair against my cheek, and the softness of her breast, warm through my thin shirt. You pay for the curiosity that drives you to these places; if you feel contempt you hate yourself, and if you feel pity it makes you miserable.

Drink was surprisingly cheap and Madame resented my unprofitable departure. The heavy door clanged behind us, and as I turned into the street five young Arabs came out from the shadows and blocked the narrow alley before and behind me. They said something to the taxi-driver, who without a glance at me (and without his fare) climbed into his car and drove away.

It is asking for trouble going into such areas alone. These gangs, whether Teddy-boys in London or Arab louts in Algiers, have little courage and, out-numbering their prey, rely on his being either drunk (because few people visit such a place sober), or conscious of a guilt complex with the fear of being found out. I wasn't drunk and couldn't have cared less who knew I was there, but care was needed—a razor slash across the arm could delay my sailing for weeks, and it was no use trying to break the circle and flee only to be caught in a blind alley.

One type came close and asked for a cigarette, and I knew my answer would decide the issue—a refusal would give the spark their attack needed, and a submissive response would boost morale. I passed the packet round, lit my own and theirs, and made a disparaging remark about the female talent adjoining. They were delighted (they may have thought I was French but my French told them I wasn't), and we all went back inside and had another drink. Later they escorted me through the maze of black alleys and stinking tunnels to the town, one got a taxi for me and all bade me farewell.

On arrival at the club I asked the taxi-driver what he wanted, and he said that it had already been arranged. So a dirty remark made the difference between a deserted corpse and an honoured guest.

And so five days passed in Bône, without having fulfilled my reason for visiting it, but other compensations had arisen to provide interest and amusement. When I returned to *Sheila* (28 November) after getting my Clearance Papers and ready to sail for Malta, I found four bottles of Algerian wine in the cockpit with a note, '*Avec sa sympathie et ses souhaits de bon voyage.*' I left French territory with regret, but richer.

8

A navigational error—arrive Malta—censorship—departure

The trip to Malta was like others in the Mediterranean, intervals of heavy rain and hail, thunder-storms, winds gusting up to gale force, and steep icy seas which smothered *Sheila*'s decks and flooded the cockpit. I had bought a set of oilskins in Algiers which proved to be quite useless, and the rain and sea between them soon had me soaked to the skin. There was no possibility of drying clothes in that weather, and so when down to the last dry set I kept it below to have something dry and warm to wear when I fed or slept. Before going above to the helm again I changed out of this dry suit into the discarded wet one, which did not matter if it got wetter. Climbing into a wet bathing suit is luxury to that.

The nights were bitterly cold in the exposed cockpit, and on going below for a hot drink after hours at the helm it was often some time before I could use my hands to light the primus. Fortunately woollen clothes give some warmth even when wet (otherwise I suppose the world's race of sheep would have become extinct long ago), and with oilskins to stop the wind I did not freeze to death.

My course lay through the La Galite Channel, in which a westerly gale rushed me north-east about twenty miles offshore and well outside the Fratelli Rocks, sail reduced to the storm jib only. I passed Ras Engels, the northernmost point of Tunisia, and crossed the Bay of Tunis via the Skerki Channel in one of the worst gales of the voyage, to arrive off Cap Bon at dawn on 1 December. I was whacked, and note from my diary that I had the sense to heave to, cook up a good curry and sleep.

Rough weather was no drawback to sleeping once I was firmly wedged in the lee bunk (the downwind side), except when a breaker came on board. Down below this sounded as if the whole ship had been smashed to pieces, and is something I never learnt to accept without a feeling of awe that such power could be survived. A really heavy sea would also fling *Sheila* with such force that I was sometimes catapulted out of the bunk to the upper far side of the cabin, to my great discomfort. Such rapid take-offs nearly always wrecked the fittings of the saloon table.

Next day the wind blew lightly from about nor'-nor'-east, and when I arrived in that bit of ocean where I expected to find Malta it wasn't even in sight. I was not really lost, because the whole of Europe lay somewhere to the north and the mass of Africa to the south, but it was an unpleasant feeling not knowing which way to turn. (The explanation of this navigational error came later during a talk with an RAF officer in Malta. I had crossed the Meridian of Greenwich from west to east, and had continued subtracting longitude from GHA instead of adding it to get the LHA.*)

Before leaving Bône I had read that Mt. Etna was in eruption, and there was a strange smell on the wind which I finally traced in memory to the chemistry lab. at school—burning sulphur. So I took the bearing of the wind and drew it on the chart through Mt. Etna, and knew I was somewhere roughly on that line; at noon I took a meridian altitude sight, giving latitude, and I was somewhere on that line—I could only be on them both where they crossed. By taking a course from that intersection to Malta on the chart and converting it to a compass bearing, I got some idea of which way to go in order to find Malta. The powerful light on Gozo showed up that night.

In Gibraltar I had asked the Royal Navy the whereabouts of the Yacht Club in Malta, because such clubs nearly always have the best anchorage and facilities for small craft like mine. It also meant that whatever hour of the day or night I arrived I'd know where to go, instead of feeling my way round a strange harbour looking for a suitable berth, or just going where I was told, which was not always satisfactory.

---

*   Greenwich Hour Angle, Local Hour Angle—Ed

This was a principle I adopted for the rest of the voyage, where feasible, and usually I wrote ahead to the yacht club concerned asking permission to berth in their domain. Over the whole voyage it was interesting to note how clubs differed, as do individuals. Some were yacht club in name but social in enthusiasm, others breathed salt spray and practised austerity in the club premises, some were wealthy, some poor, some very efficiently run, and some merely drifted on plans for the future. All, however, were most hospitable and helpful.

It was a great disappointment to learn that the closing ceremony for the season had taken place only the day before. HRH Princess Elizabeth (as she was then) and the Duke of Edinburgh had both been there, and if I hadn't lost time by making a stupid navigational error I'd certainly have met them.

Evidently I was feeling a little depressed on arrival in Malta; according to my diary written on 5 December:

> I still feel doubtful of my desire to go on with this trip. It is too hard. I would so much sooner devote all my time to thought, study and writing. Yet I know that without the hardship I'd revert to my inherent idleness—it is so much easier to dream than to put those dreams into writing. If I had what I wanted, love, security, comfort, these other things would become commonplace and their value lost. With hardship and fear the real values stand out sharply, and perhaps some day I shall find my way to them under more reasonable conditions. This trip is really a pilgrimage, a training, and I must have the guts to finish it.

One drizzly day I was driven across the island to a hill from which we gazed down to a bay protected by a long reef. It was on this reef that St. Paul was wrecked nearly 2000 years ago on his way to Rome. I later looked up the account in the Acts of the Apostles, and it could almost be taken verbatim into the columns of a modern yachting magazine. After being swept towards the shore on a dreadful night of storm, having done all they could to save the ship and themselves, they threw out anchors and prayed for the dawn. That longing for the dawn!

Each day in Malta brought an increased feeling of dejection, repression, hard to describe. At first I thought it was the usual fatigue after a voyage, and then put it down to the lack of laughter. I don't remember seeing anyone in the streets laughing, and the clothes particularly of the women were drab, waiters and shop people were sombre, and after French vivacity this was a noticeable contrast. A casual question before broadcasting gave me a lead to the answer; when seated in front of the microphone I had asked:

'Is this direct to the listeners or recorded?'

'Recorded, it has to be.'

'Why?'

'It has to be censored before we can broadcast. The Church has a priest attached to our office who censors everything before it goes over.'

I later made further enquiries elsewhere, and was told that the Roman Catholic Church censors all branches of school education, the press, the cinema, the radio, and rules even what dress will be worn in the streets and on the beaches.

Yet in Malta was one of the most flagrant honky-tonk areas seen in any port of my voyage, almost in the centre of the town. I only passed through it at two o'clock in the afternoon when it was bad enough. There was also a legalised homosexual night-club where the floor-shows were mainly female impersonation, and male prostitutes openly solicited customers as women do in an ordinary brothel. It was very well run (I was told), the 'girls' all wore a bracelet bearing their registered number, so that if a client had any trouble he could report the offender to the authorities! This tolerated perversion in the heart of enforced purity I found very hard to understand, and put it down to the extremes which are found in any dictatorship; because whether a dictatorship is economic, political, parental or religious it is still dictatorship and must bear its fruits.

I sailed out of Malta a day before my 36th birthday and four days before Christmas; 2500 miles lay between me and Aden, and I had to get there by February to catch the north-east monsoon across the Indian Ocean. Many kind people asked me to stay for Christmas, which would probably have extended into the New Year also, but I could not afford the time. And Christmas only means anything to me spent in my own home or with children. In foreign places where home life is under a disadvantage Christmas is too likely to develop into a big drunk, and I hate an occasion where you have to be gay, have to drink, and am a wet blanket to any such party.

9

An escapist—stormy weather—a floppy mast—arrive Derna

The last few days before sailing were always a frantic rush, people continually asking me when I was sailing so that they could come down and see me off. It is almost impossible to give an accurate forecast of this, because all arrangements are inter-connected and if one goes wrong it throws out everything else.

Next day, 22 December, being my 36th birthday, was the first in the second half of my allotted span; the outlook was not good. At a time when most men of my age have a family and home of their own, are secure in their career or profession, have a circle of settled friendships, I had nothing because these had either been thrown away or taken from me.

From the Industrial Revolution onwards individuals have suffered, even died, and groups have suffered ridicule and social ostracism to gain shorter hours, better living conditions, higher wages and more security. Now we have gone further and worship these under the name of a higher standard of living. Yet I had turned my back on such of this as I had had, and deliberately sought a life of very long hours, little security financial or physical, living conditions not always luxurious, and no pay at all.

As I saw it, it was not so much a sense of values as a priority of values—both gold and honesty have value, but it makes all the difference to your own and others' happiness as to which is placed first. Yet these thoughts sometimes brought a fear that I was a failure, that having thrown away so much for such an impracticable way of life was madness.

One young woman I met told me so in no uncertain terms. It was at a cocktail party and she was connected with some diplomatic or UN mission—very efficient, ambitious, a fanatical believer in the rightness of the work she was doing, and conscious of her ability to compete with men on their own ground. She even spoke like a man, but was more direct behind her privilege of being a woman.

I had only just arrived in port, and several people were clustered round expressing congratulations and envy of my achievement. The girl approached and said pleasantly, 'I'm not going to compliment you at all, but will say what I really think. I think it's disgraceful in times like these that a person with your training, education and health should go selfishly wandering uselessly round the world enjoying himself. It is sheer escapism.'

That is certainly a point of view, and it was not the remark that made me want to hit back but the attitude it implied—of a person who knows not only what is right for him or herself, but for everyone else as well; and it breaks my heart to see a woman trying to be a man, because however efficient she may be as a careerist she is depriving the world and herself of the very reason why she is in it. I asked, 'If you were sitting on a stove which got hotter and hotter, what would you do?'

'Get off it, of course.'

'That,' I said, 'is escapism.'

The weather became more erratic and squally over the next two days, the glass dropped, and high streaks of cirrus cloud ('mares' tails') and cirro-cumulus brought warning of coming wind. By Christmas Eve I was again under storm canvas, feeling apprehensive and on edge, and only sleeping from three in the morning to just before dawn on Christmas Day. My diary records little cause for jollity:

Christmas Day started badly—I got up just before dawn to make tea and found that the primus had been drenched—I'd carelessly left the hatch open last night and a crest had come below. Dried it out and started it going under a potential cup of tea—a good enough drink to start Christmas or any other great day. Later, sitting on the lee bunk drinking it, a breaker caught *Sheila* and flung me right across the cabin, smashing the saloon table—which will make chart work more difficult until I can mend it. Cut my leg.

Breakfast of boiled eggs, not because I like them but because they are easier to control during eating—I wedged my bottom well into the galley sink, braced one leg against the companion, the other against the galley shelf, pressing my back into the corner of the bulkhead—this leaves both hands free. Under my right leg is the cutlery drawer with its racks, etc., and in this I place my mug of tea, pepper and salt, spare egg, and so all is secure and to hand.

After breakfast, very tired after the lousy night, I went on deck and decided that *Sheila* could bear the storm trysail—hauled on the halyard to hear a rending noise, had forgotten (typical when tired) to undo the lashing round the sail and so tore a great two-sided rent in it. I spent the rest of the morning mending this, clinging on to avoid being thrown overboard and whipping a stitch through in passing moments of equilibrium. About every half-hour a hailstorm came over—one was so thick it filled the patch I was sewing with stones and had to be emptied several times. Too rough for sights, bitterly cold —just what is it drives a man to put up with bloodiness like this?

By next day it was blowing a full gale, roaring out of the Gulf of Sirte and building up huge seas. *Sheila* was stripped of all canvas and ran under bare poles, which was dangerous in the steep following seas, but time was getting short if I was to catch the north-east monsoon across the Arabian Sea.

A good cruising yacht, sound in hull and rigging and not carrying too much sail, well battened down, has little to fear from storms well out at sea away from land—except breaking seas. Mere broken crests may roar down on a ship, slide over the decks and fill the self-draining cockpit, and even slew the ship off course, but these are not really dangerous. Once in hundreds a wave arrives at the ship ready to break—

the high curling crest towers above and then tons of solid water crash straight down on to the decks, putting a tremendous strain on beams, deckhouse, hatches and coamings. If any part of the deck were smashed in water would pour below to fill the ship and sink it.

Breakers usually go in patches at sea, and when running before a gale these areas can be seen by day and course altered to avoid them. At night the broken water often throws a phosphorescent glow into the darkness of the sky, like the loom of a lighthouse just over the horizon (and is easily mistaken for it); again, even above the deluge of noise that a storm creates around a sailing ship at sea, the roar of heavy breakers can be heard in time to alter course and avoid them. But sometimes I was caught—the first wave always broke just short of the stern, the third under the ship; it was the middle one that held the danger.

During that night as I sat lashed at the helm, there came again that ear-splitting crash astern, then the flood of broken hissing water over the after-deck. Thereafter all noise seemed to stop but I knew that behind me, rising high in the darkness, was a great crest about to smash down over the ship. A curling wave makes no noise until it breaks, and the silence brings an added sense of menace. Then suddenly the world was lost in noise, and I was torn from the helm by that first impact of solid rushing water, twisted and spun across the cockpit. I lost all sense of direction, and time stood still in the confusion of blind darkness. It was chaos. Slowly my mind recovered… I could feel the bite of the rope around my waist, my body was pressed down over the bridge deck, as by the heavy soft nebulous form of a warm, clutching octopus (although the water was cold, it initially felt warm on my wind-chilled body). I needed air, yet there was something strangely sensual in the soft embrace, the darkness, the warm fondling of my body by the confused streams of water; there was a strange temptation to stay where I was, to submit, to revert back to death—or is it to life? I wonder if a child, the mind hardly formed, feels like this before emerging from its mother's womb?

I forced the thought and the water away, rising to my feet, one hand on the rope and one gripping the coaming, pulling myself back knee-deep across the flooded cockpit to gain control of the helm and steady *Sheila* as she rose to the third sea. She was racing forward at terrific speed, the crest lifting her high, curling above the decks on either side and then breaking with a roar of angry defeat beneath her. *Sheila* sank into the surprisingly calm pasture of smooth, gently heaving foam left by the fury of the broken wave. All other noise was stilled, as the happy tinkling sound of water was heard escaping through the scuppers back to the sea, and the hiss of millions of tiny air bubbles rising to the foam-flecked surface. I have never understood how it is that these small sounds stand out so sharply, obliterating the noise that yet must continue in the storm around them, but it is always so.

Slowly the cockpit drained, the warmth of the deluge left me as the wind's bitter fingers crept into my sodden clothing until the cold became beyond endurance. I lashed the helm hard over, and leaving *Sheila* to drift in her own care went below to change into the last 'dry' suit of clothes. If these got soaked I'd simply have to sleep in them as well. In the meantime they did give enough warmth to enable me to stay at the helm for the rest of the night—if a heavy breaker caught *Sheila* unattended it might slew her broadside on to the next, which instead of carrying her forward and breaking beneath her might break full across her and carry away the cabin-top.

Just before dawn there came a strange feel to the ship, so vague I ignored it to the extent of not seeking its cause. It was more like the mental uneasiness you accept due to fatigue or apprehension, and wait for it to pass rather than take constructive action to remove it. As dawn came slowly over the grey waste of heaving waters and flying spume I saw that the mast was strangely unstable, and that the port stays were one moment horribly slack and the next springing to bar-tautness as the mast swayed to starboard.

I lashed the helm down, bringing the wind pressure on to the starboard stays, and went forward to inspect. The rigging screws were fast and there was no reason for the slackness at deck level. As far as I could see the splices at the hounds were unaffected; but it became obvious that whatever the reason, it was aloft, and it was no use taking up the slack of the stays until I knew what caused that slackness. An increased tension might only increase the damage and endanger the mast even more. Nor was it only the mast that was in danger—with the stays slackened the deck was taking the strain of holding the mast upright, and the uncontrolled pressures of a solid forty-foot spar could tear the ship apart.

This was a wretched situation—after the days of storm and the sleepless misery of the preceding night I was in no fit shape to go aloft in those conditions of wind and sea. Nor was I sure that my added weight high up the mast would not make just the difference between it staying upright or crashing overboard.

On occasions like this, no matter how pressing the problem, a few moments' calm reflection before acting often saves much time and unnecessary danger. My greatest danger in going aloft would not be the wide swings of the mast so much as the exaggerated jerk at the end of those swings, the whip in the poorly stayed spar acting like a catapult to fling me far out to sea.

First I struggled with the storm jib, setting it backed against the wind, and although this exerted more pressure on the already overtaxed starboard stays and heeled *Sheila* further over, it did steady the mast to some extent. Then I stripped off my oilskins, pullovers, shoes and trousers, partly to allow greater agility, and partly to increase my chances of regaining *Sheila* should I be flung overboard or the mast give way.

I climbed on to the boom, gripped the mast with my legs, the halyards with my right hand and a lower stay with my left, which gave me some leverage against the forces tending to spin me round and round the mast. On that first attempt, with doubt in my mind, I got half-way up to the cross-trees and then came down defeated. My strength had ebbed fast, and perhaps I had not warmed up; I knew damn well I was afraid.

That inspection aloft had to be done—the possibility of disaster in making it was less than the certainty of disaster if I shirked it.

I climbed on to the boom a second time and by pin-pointing my mind into my hands, concentrating on my grip and nothing else, I hauled myself up to the cross-trees, stayed there while I made a careful inspection, and before starting down took a hasty glance at *Sheila* below. From that height she looked tiny, almost hidden by sheets of spray; it was a scene of undaunted gallantry and brought a wild surge of exhilaration to my precarious perch. All our best sports hold an element of danger, and so the danger of clinging to a floppy mast thirty feet above a small ship battling with a gale also has its moments. This exhilaration killed all fear, and I hastily banished it—it is this exhilaration which makes men pick up a Bren and charge a battalion of enemy, not caring whether they live or die. It wins VCs but I had to get myself home; in fear and the power of concentration it gave lay my best hope of doing so.

The inspection had shown that the splices were perfectly sound, and the slackness was caused by a missing two-inch block of wood. I had noticed and asked about this block of wood when I had taken over *Sheila* in England. It was explained that on re-rigging the stays had been slightly too long, and could not be taken up enough on the rigging screws, so a two-inch block of wood had been placed above the hounds to take up the slack. I had received a definite assurance that this was adequate and so had thought of it no more—how could I, who knew nothing of yachts, their stresses and strains, doubt the advice of an experienced yachtsman, a qualified naval architect, and a man whose profession it was to design, build and rig yachts? It amazes me now how such a make-shift job had lasted as long as it did. The block of wood had torn out, the wire loop of the stays had slid on to the hounds, and this had caused the slack.

I took up what little slack was possible on the rigging screws, then put a lashing round the stays about eight feet above the deck, and joined that with a block and tackle to a rope made fast to the mooring cleat astern. Then I hauled tight on the tackle. This took up the slack on the stays, bringing the support they thus gave to the mast slightly aft, but it was effective. I altered course to the south-east, across the wind, to keep as steady a pressure as possible on the starboard stays.

The next job was to clear the mess aft. The heavy breaker during the night had smashed the mizzen boom and torn the sail to shreds, as if it had been slashed with a sharp knife. And my excellent No. 2 jib, which had been furled in the bottom of the cockpit, had simply disappeared.

The day passed with the storm unabated and an increasing pain in my stomach. While aloft, however tightly you grip the mast, you always follow behind it during its swift arcs, and when it snaps back at the end of each sway it meets you still going in the original direction. This means, quite literally, that every few seconds you are hit in the lower portion of the stomach with a forty-foot pole, treatment no human stomach is designed to endure. Next morning after a horrible night I was bleeding inside, whereafter fatigue and anxiety did its job. Whenever I pulled a rope or clung against the violent movement, I felt the blood well warmly inside me; or imagined so.

The glass had been steadily dropping for the last three days and continued to drop, indicating that the storm had yet to reach its climax and stabilise, and then probably die at the same rate that it had risen. I could expect probably four more days of gale conditions, and if my internal injury were serious it would have little chance of healing.

That morning a big tanker passed, her bows plunging heavily into the huge seas and making it difficult to read her name; it was something like *Cabra*. My decision to signal for help was undoubtedly biased by great fatigue and worry about my injury, and the possibility of the stays breaking loose again, but luckily from past experience I knew this. I actually got out the necessary flags and prepared them to make the signal—then swore a healthy round oath and put them away. The only help the tanker could offer would be to take me off and after a few days in a comfortable bunk I'd probably fully recover—to realise that I'd left *Sheila* to her fate quite unnecessarily and never forgive myself.

The next half-hour I spent below poring over the chart, trying to work out another plan feasible under those conditions of wind and sea. I could have turned down-wind and been in Crete within twenty-four hours, but it looked a terribly dangerous approach in thick weather; and I also wanted the wind on the beam, to fill the jib and thus have a steadying effect on the mast. The nearest place I could make on the African coast was Derna, in Cyrenaica,* about 100 miles away across wind.

I took what sleep I could in snatches during daylight; my clothes, the last dry set, were again soaked, so I slumbered fitfully in these under the meagre comfort of damp blankets, awakening each time almost dopey from the sodden warmth. That night the cold in the cockpit became unendurable in the wet clothes, reaching inwards with icy fingers to hold the hours still with its intensity. At ten o'clock at night, after being at the helm since supper at seven, I looked at my watch and said: 'Only two hours till midnight when I'll go below for a cup of hot coffee. I won't look at my watch again for another hour.'

And after an age, allowing for the dragging time, I looked at my watch hoping to see eleven o'clock; it was only twenty past ten. Once you start looking at your watch time is endless, and the best cure I found was to sing—anything and everything you can remember, over and over again, plus impromptu compositions.

At midnight I went below and dragged out my suitcase from its heavy canvas wrapping in the fo'c'sle. In this were packed my best shore-going clothes, but I could endure the cold no longer. It finally came to the choice between a London suit or evening dress, and as my social commitments after dark can usually be attended in a suit but my daily commitments cannot be met in evening dress, I now chose evening dress. In the early hours of 31 December, sleepy but comparatively warm and dry, I picked up the lights on the Cyrenaican coast and entered Derna before noon in full evening dress, less tie.

---

* Eastern Libya—Ed

10

A New Year's Eve Ball—Sheila's danger—Tobruk—on to Egypt

Derna harbour was merely an indentation in the low barren coast made by the mouth of a wadi entering the sea from the desert. A breakwater curved across the indent, and a couple of wrecks awash partially closed the gap between the end of the mole and the shore, to form a fairly sheltered anchorage. An Arab pilot came out in a row-boat and guided *Sheila* inside, where I dropped anchor in three fathoms.

The afternoon passed in meeting the various officials, lately appointed pending the granting of independence to the new State of Cyrenaica. Officials of newly independent countries are always the most difficult to deal with, partly because they have little experience, and partly because they are full of nationalism and determined to show their authority. I imagine they felt exactly as I did as a newly commissioned officer taking over my first parade—very aware of my lack of knowledge but determined not to show it, afraid that my authority would not be respected and quite at a loss to know what to do if it was. However, in the case of Derna matters were simplified by the friendliness of the officials themselves, and the helpful guidance of the popular British Liaison Officer.

I was advised to engage the services of a local as night watchman, not because he would either stay awake or even see *Sheila* from his shelter on the mole, but because by paying him for doing nothing the local light-fingered fraternity would be satisfied. It was a sort of unofficial harbour due. The Arab engaged was a wicked-looking character named Ali and he had one leg, the result I learned of his previous profession of metal scrap collecting.

This scrap-collecting appeared to be the main industry of North

Africa. The locals with old trucks, donkeys, and camels roamed the war-littered deserts for metal. In their wanderings some trod on landmines, some found metal in pieces too vast for their particular type of transport—such as a burnt-out truck, a shattered tank, or a deserted gun. Their technique then was to find an unexploded aerial bomb, place it against their quarry and light a fire around it. The resulting explosion converted the prize into more readily transportable pieces, but sometimes killed, sometimes maimed, the too eager Arabs awaiting the spoil.

During the afternoon a *Girl Pat** type of vessel came in, and I hauled short on *Sheila*'s anchor to give her room. As there was still about ten fathoms of chain out in a depth of three fathoms, I left *Sheila* anchored thus instead of anchoring again further out on more chain. It was a sheltered anchorage after all.

That night I attended the New Year's Eve Ball. Several wives kissed the lone voyager at midnight and the gaiety increased until two o'clock in the morning. Outside, unperceived, the wind rose screaming out of the desert, raising sheets of spray in the harbour and *Sheila* was dragging her anchor on the shortened chain. When I arrived in the harbour her stern was only ten feet from the breakwater.

My one-legged watchman appeared out of the darkness, and explained that as soon as he had noticed *Sheila* dragging he had collected some of the fraternity, borrowed a spare anchor from somewhere and laid it out in time to save her. She was still too close to the jagged breakwater for safety, so I laid out a long 3in. warp to a mooring buoy and hauled her out to deeper water. This would not have been possible alone in that wind, but the skipper of the *Girl Pat* vessel helped me save what was nearly lost through carelessness. This lesson stayed with me for the rest of the voyage.

In the morning I expected the whole port to descend on me, led by Ali and demanding exorbitant sums of *baksheesh*. Neither Ali nor any-

---

* *Girl Pat* was a small fishing trawler, based at the Lincolnshire port of Grimsby, that in 1936 was the subject of a media sensation when its captain took it on an unauthorised transatlantic voyage.—Ed

one else came near me, and only by seeking him out with the aid of the harbourmaster was I able to trace my benefactors.

Such local help came in various forms at most eastern ports visited. It seldom took long for the story of *Sheila*'s voyage to go round a port, and the Arabs, who have sailed from the Mediterranean to Africa, to India, and to the Java Sea and beyond for centuries, paid their respects in whatever way they could.

After a week in Derna I sailed eighty miles further along the coast to Tobruk, and over the next three weeks of cold, biting winds I made the necessary repairs to *Sheila*'s rigging.

Tobruk was a dump. Wrecks still littered the harbour, the rusted masts leaning above water to mark their graves as though forlornly begging remembrance. The town was torn and scarred, desolated by great bared areas where debris had been cleared away. I did not visit the cemeteries, large fields of neatly planted crosses, because I can remember what the dead gave. A man is not loved for his flesh and blood so why weep over these when the rest has flown?

It blew quite hard on the morning of 21 January when I sailed out of Tobruk under two reefs headed for Port Said, 500 miles along the North African coast to the east. As always on leaving port, I felt 'butterflies in my tummy', particularly as darkness fell. I'd had no food all day, having missed breakfast in the rush of preparing for departure, and lunch while leaving harbour, and so only after dark did I cook up a good stew. The smell of cooking food made me retch, but this nausea passed by midnight when I had a good hot meal and slept the rest of the night in hour shifts.

The weather during that voyage brought no storms and was for the most part gentle, too gentle at times. The loss of the mizzen told particularly in light winds, but when these were too strong for the big No. 1 jib (having lost the smaller No. 2), the storm jib was too small. And three days out the wire luff rope of the mainsail broke just below the first reef cringle, and I was unable to repair this without fear of the rotten sail tearing badly, and so completed the voyage with one reef in the mainsail and further loss of speed.

One night while changing jibs I snagged the halyards aloft, and rather than climb the mast in the darkness, I played a torch on them from the deck to trace the offending rope. It was difficult to hold the torch with one hand, free the guilty rope with the other, and yet maintain balance on the plunging deck.

A huge liner came up from astern and passed very close, and some intelligent officer on the bridge must have guessed my plight. A powerful searchlight blazed out, lighting up the mast and ropes like daylight, and stayed there until by its aid I cleared the mess. I am grateful to that officer, whoever he may be.

The wind steadied by two o'clock that morning, so I left *Sheila* to sail herself on a course just south of east while I went below to sleep, intending to awake at dawn to get star sights. But my body knew best and held me in the sleep I needed until a voice awoke me at 7.30—that same voice I had heard before and was to hear again.

'You'd better check that helmsman of yours. He's off course.'

I arose refreshed, and going to the helm noticed that the morning sun was astern. The wind had turned from north-east to north-west, turning *Sheila* with it and was sailing her back towards Tobruk.

For several days after that *Sheila* lay becalmed. I escaped the weight of physical inactivity by wandering in mental fields.

There were many books in *Sheila*'s library, but before leaving England I had added others I knew I would need. My main initial interest was philosophy, but this had shown its interconnection with religion—with all religions because all seek the same thing, whatever each may conceive God to be or by whatever path they seek Him. It had also led me to study art, because that also seeks Reality, the goal of philosophy itself. Environment and history affect men's outlook and sense of perception, as does their education, their form of government, their economic system, and so my search for knowledge broadened into many fields. And in these was science also, because science too is Truth. It is a very different way to that of mystical thought (to which I was more inclined), but the mystical thinker cannot deny science or he denies Truth itself, his own goal; nor should science deny the mystic.

The Hindus liken Truth to the topmost peak of a mountain, which can be reached by many different paths from all sides. Although all those paths may differ greatly in distance and grade, in smoothness and hazards, the closer they get to the top the more they come within each other's view; near the top some even merge into one, although they started from opposite sides of the mountain. At the top they all become one, having reached the common goal.

And so I had bought myself a book on psychology and read it during days becalmed. It was an elementary book written by a man who was certain of his path towards the answer, but he made the not uncommon mistake of also believing that his path was the only path. To induce an equal belief in others (with the best of intentions), he wrote to the effect that although psychology is a new science, in which cause and effect cannot be so easily demonstrated as in physical science, it is none the less a science in that it employs scientific methods of research; and the conclusions reached are thus scientific fact.

I could not accept this lure to conviction. For instance, if a man stands at the head of a light railway facing a mountain of clay, equipped with pick and shovel, gelignite and so on, and digs a tunnel scrupulously employing coal-mining methods, he won't get coal.

And not once in that book did I find mention of soul, spirit, or conscience, nor anything connected with these. Whether a psychologist believes in the existence of these or not, the majority of people (his material) do, and their behaviour is greatly influenced by that belief. And so surely a psychologist is forced at least to mention the spirit, but there was only much talk of the subconscious. These two points annoyed me so much, coming from a supposedly intelligent person, that I threw the book overboard.

Months later in the Red Sea, thinking about Darwin and the Church, my thoughts returned to that book. Both scientist and Church agree that we are made up of three parts, and agree that the first two are mind and body. The third part the Church calls the soul, and science calls the subconscious—but is it possible, I wondered, that they are talking about the same thing, called only by different names and seen from a different angle and with a different outlook? My further studies in these vast subjects have led me to consider this might be so; a quibble in words, but what damage has it done?

II

Arrive Port Said—Mr. MacGregor—departure from Port Said

My one idea on arrival in Port Said was to get out of it as soon as possible, to continue my voyage and reach Aden in time to cross the Arabian Sea to India with the north-east monsoon. But first I had to renew *Sheila*'s rigging for the dangerous Red Sea passage, and collect the new mainsail which I hoped would be waiting for me.

No officials visited me the first evening, so next day I left *Sheila*'s flags flying to indicate that the ship still required pratique. The morning after that I rang the Port Authorities who arrived shortly after, very annoyed with me for having been ashore before obtaining clearance-in. I explained that it was the authorities' responsibility to meet an entering ship flying the necessary signals, and as these had been flying on *Sheila* for two days without response I had had to go ashore and telephone to draw their attention to this fact. There was no reply to that.

The Egyptian police were very suspicious, and continually asked a question signifying a suspicion which I found prevailed in official minds in every port from Gibraltar to Australia:

'Why are you travelling alone?'

Is it possible that the type of person who values a government job and the security it gives finds it difficult to understand anyone who forsakes security, and finds it impossible to understand that anything else exists which is more valuable than security?

'Why,' they asked, 'when you can travel in more safety and comfort, and even cheaper, do you choose to travel like this alone?' The reward to their minds could only lie in something illegal—dope-peddling, gun-running, spying, for instance.

About five times during my fortnight's stay the police launch ran alongside *Sheila* where she was anchored off the *Cercle Nautique*, and ordered me to police HQ. There the ritual was always the same—about an hour's wait in a room full of junior officers, who gave me coffee and opened a friendly conversation which veiled searching questions of my past experiences and future intentions.

Next I was shown into the head office where a tubby little Lieut. Colonel in the five different interviews tried five different types of investigation; coolly official, keen friendly interest, detailed investigation, threatening, and finally open accusation ending in, 'Are you spying for the British Army?'

As soon as possible I went to the bank to collect mail. No package (the sail) awaited me, and nor was any letter from the sailmaker in my mail. I spent a valuable pound in a cable asking if the sail was en route.

The next week brought no reply and so I sent another cable, this time reply-paid, and again valuable days passed while I lived as cheaply as possible. My main diet was Arab bread bought in the bazaar (it was cheaper than European), and sometimes for variety I visited a cafe and ordered a small glass of beer. This entitled me to a free meal of the plentiful *hors d'oeuvres* littering most Mediterranean bars… olives, shrimps, cheese, octopus, gherkins, onions, and so on. Eventually the reply to the cable came, merely saying that the sail had not yet left England.

On one of my earliest visits ashore a cheerful gentleman in a red fez introduced himself as Mr. MacGregor. He then made the most astounding suggestion, even for Port Said; it was astounding not so much for its depravity but because it was made at all. Who would respond, I wondered, and determined to find out.

I laughed and declined. Mr. MacGregor looked at me quietly and said, 'Sure, you no want. You no get angry.' And after that on other visits Mr. MacGregor always met me to talk of this and that, and we became quite friends. One day I asked him to join me in a glass of beer; he was a dreadful-looking person but travelling alone I had no social conventions to maintain except my own. After some small talk I asked Mr. MacGregor to whom he sold his 'feelthy pictures' (which was only one of his interests, and not the particular depravity mentioned above).

'Oh, when ship come in.'

'And when there's no ship, to rich Egyptians?'

'No,' he said. 'Egyptian no buy. Only good man from ship buy.' (What on earth did he mean by 'good'?) And then being a born mimic he put on a little play for my benefit.

'This morning time,' he said, meaning Scene I. He portrayed a respectable European official strolling through the crowds ashore with his family. On being offered 'feelthy pictures' he became a blaze of outraged dignity, whirling his cane and calling Mr. MacGregor a 'feelthy beests'.

'This afternoon time, when everybody sleep,' (Scene 2).

Again this character strolled between the cafe tables (almost deserted, thank goodness), this time as a furtive individual acting casual curiosity in a strange land.

'Psst. You got postcards?'

Here Mr. MacGregor became himself again, and handed over a discreetly closed envelope ('These very good pictures, Sir'), asking an exorbitant price. This the respectable customer paid without demur, only longing to end the humiliating situation in which he might be caught.

I could remember as a young subaltern passing through Port Said, the morbid curiosity reaching far back into a dark recess of the mind desiring to see those photographs. I had approached several vendors in the crowded streets but had lost my nerve to buy any, although the sense of guilt had destroyed my happiness just as effectively as if I had.

That evening in the ship's lounge four of my Army friends sat round a table from which came sudden outbursts of barnyard noises. I wandered over.

'Get out your photos ol' boy and join us. We're playing animal grab.'

I watched amazed—each that day had bought a packet of 'feelthy pictures', and now in turn dealt them one by one face up on to a pile before himself. Each player too was a pre-arranged animal—a cock, a donkey, a cow, and a pig.

Charles played a card depicting one aspect of the erotic game, and all eyes hastily glanced at other stacks to see if it was duplicated. It was not, and so the game proceeded in silence. Then Bill played, and although viewed from a different angle it was the same as Charles', who noted that fact first and immediately brayed loudly. Slower minds laid claim in the determined grunting of a pig, the urgent mooing of a cow, and the strident crowing of a cock. Charles' claim was sustained and he reached to gather Bill's photographs into his own.

Yet around that table of extraordinary noises and laughter the mass of photographs lost their evil fascination; openly, they were seen in the ridicule of their grotesque indignity to lighten that dark recess.

Emile was one of the first people I had met in the *Cercle Nautique*. He had some job in the canal company, as far as I know not very important, but he loved the sea and ships and his help was always unobtrusive. We were sitting on *Sheila* talking and he asked the condition of the sails.

'They will need a lot of sewing,' I replied, and within a few days found two balls of first grade sewing twine left in the cockpit.

He also arranged for technicians to do various electrical and mechanical jobs for me. One of them was a huge half-Egyptian, half-negro, who spent several hours on board after his day's work. He spoke no English, but we got along in French. He took infinite trouble to do a really good job and when it was completed I asked what he wanted. Looking at me steadily he said quite simply, '*Pas de rien Monsieur; et pour Monsieur Emile—il est tres gentil.*' Everyone I met had a respect and affection for Emile and I owe much to his kindness.

One evening I was travelling on the ferry between Port Said and Port Fuad with two heavy coils of wire rope. I was reasonably well dressed in a tropical suit, and it must have looked strange in that country to see a white man carrying a coolie's load. Opposite me sat a young, neatly-dressed Egyptian, and from his expression it was very obvious that that was just what he was thinking. As I caught his eye I couldn't help smiling.

When the ferry stopped the Egyptian came over, picked up both coils, and asked me where I was going. At the Customs barrier I said, 'You'd better leave me here. I've got my receipts but this will take hours because I won't give a bribe.'

He looked at me a moment, said something to the official at the barrier, and on we went to place the wire on board. He shook hands and left, and this in the days when considerable anti-English feeling was fermenting.

I went to the canal company office to ask permission to pass through the canal. The French officials gave me a huge form to fill in, whereafter they perused it carefully. On reaching the item 'Number of Crew— 1', they said, 'Monsieur please, a mistake?'

So I explained.

'A-a-ah, Alain Gerbault, yes?' Alain Gerbault* is still something of a national hero to France, and after a lot of discussion in the manager's office they returned to say, 'For you, monsieur, no charge.' (I could not have gone if there had been!) And nor did I have to take a pilot.

And so the evening before the day of departure came. The steward (a Lebanese) of the *Cercle Nautique* had obtained all my stores for me, saying that he would get better quality at less cost than I. I had ordered no cigarettes and with a sick feeling asked for the bill. I gave him all I had, the last of my non-sterling funds, and it was three piastres too little.

'No matter,' he said. 'It is nothing.'

It is nothing—so many people said that to me in foreign lands, but without that nothing, my trip would never have been completed.

---

* Alain Gerbault circumnavigated solo in the 1920s in *Firecrest*—Ed

12

*The Suez Canal—French women—arrive Suez—
a veiled proposal—clearance-out*

The dawn on 17 February, 1951, was still and clear as I took in *Sheila*'s anchor, and motored through Port Said harbour to enter the canal to Suez.

Certain incidents of this voyage, apparently so undramatic and unimportant, stand out clearly in my memory, and this is one of them—the number of times I had already passed through this canal in the past. First as a youngster straight out of the New Zealand bush on my way to Sandhurst, then as a subaltern going on leave, and the last occasion as a fully fledged Major with war service. On all those occasions I had felt the thrill of passing through that huge man-made waterway connecting East and West, but on no occasion did it enter my mind that I would one day take my own ship through it. And which was the dream—those moments when *Sheila* chugged through the glass-like water, or those memories of a life so different?

Now too was an expectancy; behind lay the Mediterranean with its bitter winter seas, and the nations I generally accepted as being 'civilised', but before me lay the Tropics and all the fascination of the East with its different customs, religions, and 'lesser breeds without the law'. Here my voyage in a sense began, like turning over the introductory page of an adventure story.

The canal authority had explicitly laid down two restrictions. The first was not to travel during darkness, as *Sheila* was not equipped with the searchlight prescribed by canal regulations. As a corollary to this I could only tie up for the night at a recognised staging post. If I tied up just anywhere along the bank I'd be shot without warning as a dope-smuggler. My plan therefore was to reach Ismailia before dark, a distance of forty-two miles.

The second restriction was that when passing convoys I must leave the main buoyed channel clear, and move into the water between the buoys and the bank. '*Attention, Monsieur,*' the French had said, explaining that as the huge ocean-going ships forged through the confined channel they pushed a great volume of water ahead of them. It mounted high along the banks until it escaped past either side of the ship in a torrent, leaving the shallows of the banks high and dry; if *Sheila* was too close to them she also would be left high and dry.

I headed straight down the centre of the buoyed channel, the ribbon of glass-still water stretching out of sight to the first slight bend 25 miles away in the desert. The mounting morning sun soon brought a shimmering haze, and the mirage effect of sandy hillocks standing isolated in mid-air. The smoke of the first approaching convoy showed as a smudge in the haze, then came the masts, the superstructures, and finally the majestic hull of the leading ship approaching with cautious surety, proudly uniting the two worlds through this wonder that man had created with his own hands. This canal is indeed a miracle '—if you believe.' De Lesseps had believed and the canal was built.

I turned aside from the main channel when the first ship was still two miles away, partly to get the feel of the shallower water before the test of passing came, and partly to leave the channel clear to the pilot's eye. As the ship drew closer, towering above the desert on either side, I could see the hump of piled-up water mounted before her bulk. It came clutching along the sandy banks and then sped past the ship's flanks, sucking away from the banks to fill the space the moving hull vacated.

Sheila felt the movement, seemed to climb slowly, and for a moment the high flare of the steamer's bows were reaching over me, inescapable, but I resisted the impulse to sheer away towards the bank which was already close. Suddenly we slipped downhill, shot swiftly past an ironclad wall on one side and the sucking, gurgling bank on the other. Such moments bring a thrill in handling a small ship.

And so the day passed. The scenery changed little, the long dead-straight stretches of mirror-like ribbon glared monotonously ahead, and the sun turned the open cockpit into a thirsty oven. It was with relief in late evening that I turned out of the main channel (where the canal continues on through Lake Timsah), and headed for the settlement of Ismailia. But first a French yacht was waiting to meet me, and drew alongside to hand me a tall ice-cold glass of lager. That is civilisation. 'When are you leaving?' they asked.

'At dawn tomorrow,' I replied, not knowing then that work and play was to keep me a week.

That evening we had a few drinks at the *Cercle Voiture*, and fed at an Egyptian cafe. I always go out of my way to eat the food of each country visited, partly out of curiosity and partly to destroy prejudice. The idea of eating frogs' legs once repelled me, but now I know that they are indeed delicious, and after all a frog is a much cleaner animal than a tame pig.

I was then taken to a dance at the French Club, and introduced to some charming mademoiselles as *Monsieur le Navigateur*, which sounds so much more interesting than Mr. Sailor. I had a few dances but was not in hilarious form; it had been a tiring day and my mind was preoccupied with the Red Sea, how and when I would be able to re-rig *Sheila* for it, or tackle it with the rotten mainsail.

But I could watch French women for ever. They are so vivacious, so very feminine, and the plainness of the few is unnoticed because they are always so beautifully turned out. It is as if they are proud of being women, fully content with the joy and power of being a woman, which after all is God's greatest gift to Man after Himself— and also perhaps His greatest rival, as Eve well guessed and celibate monks suspect.

Other women of the West seem to feel an inferiority, and so strive to kill their femininity in order to row boats faster than men, jump horses higher, throw javelins further, organise society better; and to attain those aims they have to kill the greatest hindrances to those achievements—their frailty, tenderness, beauty, things which are born into them. It is a high hurdle they set themselves, and so too often they enlist the aid of fanaticism to do it. And a fanatic is a menace anywhere.

In the early hours of the next day we moved on to another dance at the English Club. And here most unexpectedly I came in the way of getting *Sheila* re-rigged, so later told my French hosts that I would move *Sheila* over to the British Club.

'But,' they said, 'we can do everything here. You only have to ask.'

It is horrible to wave aside such an offer of help, and it must have appeared ungrateful and 'clannish'. But I had had so many offers which had come to nothing, which had raised my hopes only to leave me bereft, and I half-expected these offers in Ismailia to be the same. I could deal with such a situation better in my own language than in limited French, and I could also explain in English exactly what I wanted; few of the French understood English well enough when it came to technicalities. I simply could not afford to take any risks where it came to *Sheila's* rigging, so courtesy and apparent gratitude were two of the sacrifices made, unwisely as it happened.

And so over the next week *Sheila's* standing rigging was renewed with the wire rope I had unexpectedly found at the last minute in Port Said. A kind Englishman had given me the equivalent of £15 in Egyptian money to pay for it, and later I sent a cheque (the proceeds of an article sold in England) from my London bank to his.

The forty-four-mile passage of the canal to Suez was uneventful, and I arrived just after dark to berth alongside a wharf. It was my intention to post a last letter home first thing in the morning, and sail before the Port Authorities ever knew I'd arrived.

But I heard that the *Orion* was stopping at Suez that day, and on board was a girl I had known in New Zealand. After the past weeks of being pushed around by foreign officials and merchants, of living on Arab bread and struggling with strange tongues, I longed to meet someone from my own country again.

The *Orion* was only reachable by launch and the ticket was unobtainable without a police permit. I got the ticket and also the inevitable questions, 'Who are you?' 'Why are you here?' and 'Why didn't you report your arrival?'

Whenever I've gone on board a liner during this voyage it has disturbed me, bringing back the contrast of my previous life so vividly. Again the wide, clean lounges, the stewards to bring iced drinks, the carefree atmosphere so far from any worry or responsibility. And now in the midst of that a girl sat with me, so cool, so lovely, a friend such as had not been in my life for so long. Unfortunately too the breeze through the open glass door sometimes carried a subtle perfume from her, and perfume always upsets me. How I wished the voyage was over!

On the way back in the tender the police launch drew alongside and ordered me on board. I was later taken through the darkened docks to a tiny shed and questioned abruptly, standing against a wooden wall while my interrogator talked on the phone presumably to higher authority.

'Who was that girl you met?'

'Why did you meet her?'

'Is she your wife?'

'What did you talk about?'

Finally just before midnight I was released with orders to report next morning (it was no use trying to sneak away that night as I could only go one way and a launch would easily have caught me).

I returned to *Sheila*'s tiny cabin, so small and dim after the spacious saloon, and so lonely. Such loneliness bordering on despair is never known at sea; and so I wrote a long letter to the girl's London address, and months later in Aden received her reply.

'I've read and re-read your letter,' she wrote, 'and can't decide if you were just feeling lonely, or whether it is a cautiously veiled proposal.'

At the wharf police station next morning they immediately took and retained my passport. This was quite outside their rights, but I gave it up without demur to allay their suspicions and to facilitate an unauthorised departure should they finally drive me to distraction.

From ten in the morning until four in the afternoon I trudged the dirty, dusty, fly-infected streets from one widely dispersed office to the other. At the last office that afternoon the Egyptian official said, 'Now take this form back to the other offices, show my signature, get their counter-signature, bring it back here and I shall sign your final clearance. That will have to be on Monday, because we are closed tomorrow (Friday, the Mohammedan Sunday), Saturday is a usual holiday, and Sunday (the Christian holiday).'

I went straight back to *Sheila*, motored round to the wharf police office which held my passport, and dashed into the office obviously in a hurry.

The police officer was friendly and offered me a chair and coffee (during which time he might ring the authorities to check my clearance). I declined the coffee, explaining at length about tides, anticipated winds, and reasons why I must sail at once, using as many nautical terms as possible. (People the world over prefer to pretend that they understand nautical terms, rather than admit they do not.) Anyway the officer was impressed, gave me my passport, and wished me bon voyage. I leapt aboard, cast off, and with full throttle departed into the Gulf of Suez, the 180-mile arm leading into the Red Sea proper. Suez was the first port I left without clearance papers, and I carefully shook the dust from my feet over the stern.

13

*The Gulf of Suez—planning for the Red Sea—doldrum weather—
asleep among the reefs—arrive Perim Island*

Sat. 24 Feb. '51. Off Ras Gharib, has blown up into a full gale, under storm canvas—how I hate sailing with wind and seas dead astern. Getting tired, having been at the helm since leaving Suez 30 hours ago. Reefs, shipping, confine me—would be madness to sleep.

With all my careful planning of the Red Sea passage I had tended to overlook its very threshold, the insignificant entrance through this gulf. In some ways it was the most dangerous part of all, and had finally to be treated as a phase by itself.

Its average width is about twelve miles, which meant that I could not get further away than six miles from the nearest land, and this I consider is too close to sleep safely. Reefs and shoals in some places force shipping into narrow channels which demand extremely accurate navigation, and I had as yet little reason to believe that mine was. Currents of unpredictable strength and direction complicated matters even further, and with the world's shipping concentrated into the narrow waters the chances of being run down were great. I knew that I would not be able to sleep between Suez and the Strait of Jubal, the exit into the Red Sea, and my plan was to race to clear that narrow 180 miles before the limit of my endurance ran out.

The chart showed anchorages of sorts along the coast in which I might have got sleep, but they were all tucked in behind intricate coral reefs. At that time I had no experience of coral nor the technique required to deal with it, and those were the worst conditions in which to learn. The Egyptian police had also stressed the danger of entering any coastal settlements, as policing was nil and the inhabitants were lawless. In that part of the world many Arabs are fanatical Mohammedans, who believe that they gain merit in the eyes of Allah by slitting an infidel throat. I have no quarrel with anyone's belief but in this case it was my throat.

That second night I stripped *Sheila*'s masts of all sail except the tiny storm jib, and flew down the coast struggling to concentrate on the compass course. As I was very tired, sitting in the exposed cockpit enclosed by solid darkness, the one circular light of the compass bowl had an hypnotic effect; it seemed to draw my whole being into its lighted circle, taking me out of time to dissolve me into nothingness, and demanded a consistent effort of will to resist.

Dawn broke on the wildest scene I've ever known, and perhaps being light-headed with fatigue and hunger made the impression all the more vivid. On either side were the great barren mountains, giving a sense of timelessness as if they had outlived their own souls, and in between lay *Sheila* tossed by a furiously angry sea. And above all lay a mist thicker than monsoon rain, the flying spume torn from the wave-tops.

We were fleeing headlong towards a narrow exit, and I strained my eyes urgently in the fleeting moments on each crest, trying to pick up the lighthouse; if I missed it by too much when already too close to the reefs, I would be unable to manoeuvre across the wind and in moments *Sheila* would be smashed to pieces. Each second counted.

I have a general rule for dealing with situations where each second counts, and that is to have a big mug of hot, sweet tea. It has the beneficial effect of putting time and all else in their right perspective. In this case also I knew that in the strait I would not be able to leave the helm for a moment, so I had to have some nourishment to sustain me for a long time ahead. This was not so much a matter of personal comfort as of necessity; without nourishment apprehensions tend to loom larger than they should, to influence judgments and distort plans, when a surfeit of caution can lead as close to disaster as the lack of it. I had then been forty-eight hours without sleep, and had to reckon on another twelve during which I would need all my wits about me.

I took very careful bearings from the hills, laid off a course for the entrance mark, unlashed the helm and turned down-wind.

The huge seas rose astern, gallons of solid water were torn from their crests and flung through the rigging, *Sheila* was lifted and driven forward like a surfboard, broken water flooded the after-deck, and as the wave roared along amidships the side-decks were buried up to the coamings. And above it all was the noise, a wild, magnificent, unrestrained orchestra—the wind really did scream and howl in the rigging, the bow wave gave out a throaty roar, and the seas replied in thunder.

Those moments stand out as among the most exhilarating of the whole voyage—a situation where the very magnitude of the forces arouses pure excitement beyond fear, but where you step your mind on to a higher level again, excluding youthful exhilaration and concentrating your technique not to guard against the forces but to harmonise your faculties with them. Fear is then banished not by the excitement of a gamble, but through harmony with the forces; you yourself become part of their magnificence and so invulnerable to them.

A darkness in the mist astern grew larger, and soon a big tanker loomed up to pass me to starboard. Our courses seemed to coincide, but a check from the beam can be very deceptive and to make sure I altered course so as to pass 400 yards under her stern. When her masts showed in line I checked her course again to confirm my own; she probably had the entrance mark plotted on radar, which had a vision though the mist denied my straining eyes. The tanker was soon lost in the spume, but ten minutes later from the top of a high-lifting crest I caught a glimpse of a slim shadow fine on the starboard bow. Soon I was certain: it was the lighthouse marking the entrance of Jubal Strait.

During that day the wind eased a little and I got on the storm trysail to help *Sheila* clear the bordering reefs. By dusk I was into the Red Sea proper and set a course into open water towards the Brothers Light, perched on its isolated pinnacle of rock eighty-five miles to the south. By this time, after sixty stormy, anxious hours at the helm, I was utterly whacked. My whole being cried aloud for sleep, and although there was much shipping about and it was still not safe, the need for sleep became greater than its danger. I slung a light in the rigging and slept; if I had been floating down the Styx in a punctured rubber dinghy I would have slept.

I awoke just after dawn to find a big sea still running under a near gale from the north. My body felt stiff and my mind shocked by the awakening, while yet needing more rest. At such times the idea of food is nauseating, but I needed food to carry me further—that wind was too good to miss, and being then in open water I was prepared to drive myself to the point of exhaustion while it lasted. My rations were limited and a fast passage essential.

That night the Brothers Light showed up and the wind still held, and with only the odd nap when sleep almost overcame me I continued to use it—the paradox of the sea! With the disadvantage of sailing into a head wind, if you fall asleep at the helm the ship continues on her way; in fact, you can go below to your bunk and sleep; but with the advantage of a fair wind astern even a moment of sleep may let the ship gybe, when the mainsail swings violently across possibly to break the boom or even carry away the mast. So those odd naps I had to take when the need for sleep overcame my utmost will to resist.

One hundred and ten miles south of the Brothers is the Daedalus Reef, merely another clump of coral standing isolated in the 100-mile width of the Red Sea, rising sheer from surrounding depths of 500 fathoms. Its light showed up after dark and the wind died to nothing. Even fair winds do not last for ever, and gratefully I went below with 200 miles of the passage behind me. I needed a good hot meal but was too tired to eat, so lay on my bunk only to find that I was too hungry to sleep. I dragged myself up, cooked a good meal, and slept.

The sun was well above the horizon when I awoke. *Sheila* lay motionless telling me she needed no attention, and so I continued to be in that comforting luxury when sleep has gently released the body after giving all it needs. I rose up to get a cup of coffee and returned to my bunk to luxuriate further. This was millionaire stuff.

After breakfast there was nothing to do without wind, so after cleaning up the ship and drying out clothing, I took out the five charts and the Pilot Book covering the Red Sea, and carefully reviewed all the information on which I had based plans for dealing with it. No one else had so far attempted to sail this sea single-handed.

The charts and Pilot Book, as well as Robinson's book *Deep Water and Shoal*, showed that there were two routes available. The first was down either coast inside the reefs where the water is smoother, sleep easier in the sheltered anchorages, and the risk of being run down by steamers nil.

I waived this route because of two disadvantages, the first being the risk of molestation by local inhabitants, and the second the difficulty of negotiating confined coral waters single-handed. Coral reefs and clumps often rise steeply from deep water and stop about five feet below the surface, so that they are difficult to see in time to avoid them from a position at the helm. I therefore decided to take the more open course down the centre and devoted my further studies to that.

In February and March the winds blow down the Red Sea from the north, but die half-way down in about latitude 22 north. They blow from the south in the southern half, tearing through the Straits of Bab-el-Mandeb, often enough at gale force, but they also die near the centre in about latitude 19 north. In the centre is a stretch of about 150 miles of no wind, but in this area thunder clouds continually develop and disintegrate, and under these are squalls of hurricane force. To deal with that area I had cut down on my stores account and bought twenty gallons of petrol.

The waters are fairly open in the northern half but towards the south the chart shows a mass of reefs. Very few are lighted, and most are hidden a few feet below the surface. The seas break over them in heavy weather, and when the sun is shining the colour of the water gives warning, but at night there is no warning, and the only safeguard is extremely accurate navigation.

Sun sights in the Red Sea are open to doubt owing to refraction of the sun's rays, and a doubtful sight is sometimes more worry than none at all. Star sights are not always possible because of an ill-defined horizon, veiled by haze or dust particles. DR is also unreliable on account of currents unpredictable in strength and direction. Accurate navigation seemed to be an impossibility.

Some think that there is little need to allow for currents in a small ship, but this is just where there is need. A steamer travelling at eighteen knots covers 100 miles in six hours; if there is a half-knot current on the beam it will put her three miles off course at the end of the 100 miles, but the same distance might take *Sheila* three days or seventy-two hours, in which time that same current would put her thirty-six miles off course. The currents are unpredictable in strength and direction, and so I never knew what to allow for them and sights were often too unreliable to tell me.

During the week after leaving Daedalus Reef I covered another 400 miles to the south, under light northerly winds which rose and died with the sun, and which imposed a routine way of life. I awoke at 4 a.m., had breakfast, and hoisted full sail, including the spinnaker, as the first breeze came at about 5 a.m. I then had to stay at the helm all day with the spinnaker set, until anywhere between 6 p.m. and midnight the wind died to leave me becalmed. Then down came the sails, a big evening meal, and sleep.

In these days it was strange to know again a tropical morning, when the cool of night still lingered before the warning of violent heat in the coming day. It always brought a sense of wonder, as if one lay exposed waiting for some inevitable miracle to happen, not knowing what it was to be. And with the tropical mornings came the first flying fish in latitude 24 north, and that was the first night I slept without a covering.

There were many dolphin (about eight feet long) in the Red Sea, and one school of seven followed me for hundreds of miles, calling to play for an hour or two and then disappearing for several days. In this school there were always just the seven; one had a stumpy dorsal fin with a white scar where the tip was missing, and one had a very wheezy exhalation whenever he blew. I told him to gargle in salt water. Sometimes at night when becalmed and sleeping on the side-deck instead of in the stifling cabin, I awoke to the soft explosive puff as a dolphin blew close beside the ship, and often Wheezy was among them.

This school became my special friends. Later in the southern winds when *Sheila* sailed herself into a head sea, I stood on the end of the bowsprit with the school in graceful escort round the bow. As the bowsprit plunged ankle-deep into a sea a dolphin sometimes sidled in close for me to rub my bare foot along his flank. On days when *Sheila* lay in a glass-still sea they swam close to the bowsprit and then dived straight down, fathoms down, until only a speck in the clear blue depths. There was a flash of white as they turned, and then came straight up like a rocket, leaving the water only six feet from me, shooting high into the air to curve gracefully over and land with a joyful splash.

There was only one hard blow that week, which forced me back to storm canvas. The topping lift gave, the clew cringle tore out of the storm trysail, and the jib sheet broke, all indicating that *Sheila* was beginning to suffer from the wear and tear. I was too, because the heat was terrific night or day.

After this the winds noticeably weakened and had soon died altogether by mid-afternoon. The decks by 8 a.m. were too hot for my hardened feet, and at noon, travelling under the spinnaker at almost the same speed as the wind itself, the heat was intense. One of my water tanks had leaked badly and I had placed myself on a daily ration of two pints, one of which was taken at sunrise and one at sunset.

Several times during my Army service, water had been limited to this amount, and the only way to endure such a ration without going mad is never to drink while the sun is hot (otherwise the body throws the liquid out in perspiration), to drink only at those times decided upon, and then drink very slowly. And take extra salt.

I had left Suez with only three weeks' supplies of food, not having enough cash to buy more. It is only common sense when the future is so unpredictable to hold a reserve in hand, and if you have not got a reserve then one must be created.

The only way I could create a reserve was to go on to half-rations, by having two small meals a day instead of the usual three full meals. This was not as drastic as it sounds because eating is largely habit, but changing this habit, like any other, is initially unpleasant.

Such a reserve must be built up long before the need for it arises. You can extend the last four days' rations into eight, thus allowing only four extra days to the voyage; but by similarly extending three weeks' rations you gain an extra 21 days. The only difference is that you go hungry longer, but not for ever!

On latitude 19 north I entered the belt of calms, and in this latitude also the offshore reefs increase and in places extend as far as eighty miles off the coast, which of course is out of sight. I started the engine for the hundred-mile run, but almost immediately it developed a petrol stoppage. In reassembling the carburettor I screwed a bronze nut too tight, breaking the bracket holding the jets. I tried for days to repair this with marine glues, lashings of cotton thread, splints of match-sticks, various packings, but none availed. The next 100 miles of intense heat took twelve days to cover.

These doldrum conditions demanded more work than sailing in heavy weather. I only snatched short periods of sleep so as not to miss any puff of wind, and when these did come they were very mild and very erratic. As a darkening ruffle approached over the glass-like surface I got on all sail, the spinnaker say to port, and after an hour or sometimes after one minute the wind died altogether or else changed to the beam—down spinnaker and up genoa, never knowing whether the wind would hold long enough to justify the effort but having to make it in case it did.

The thunder storms predicted in the Pilot Book also developed at any hour of the day and night. A small, isolated, cotton-wool cloud would grow within an hour into a towering cumulonimbus, and the squalls underneath were often true hurricane force. As soon as I could see the darkness of these squalls on the horizon (which from *Sheila*'s cockpit is only three or four miles away), I hastily took in all sail and had it lashed before the squall struck. It heeled *Sheila* far over even under bare poles. Nor was it possible to collect rain which sometimes came with the squalls, because the terrific wind tore gallons from the sea, and for twenty feet above the surface there was a flying cloud of salt water.

Yet these days were not unpleasant. I could do nothing about the conditions except the best within the limitations they imposed; I could not change them. This period reminded me of the Chinese proverb, 'If you've got to be raped, relax and enjoy it.'

I read a lot, wrote notes for my book, and watched with interest the surprising amount of sea life. Sometimes small striped pilot fish joined *Sheila*, usually forecasting the appearance of a shark. Some of these were bulky ten-footers which weaved from side to side astern if *Sheila* had way on, but when becalmed they often came to be right alongside in the shade. I always drove them off when possible because of the sailor's superstition that a following shark predicts a death on board. It is easier to laugh at such nonsense on a crowded liner than it is when single-handed.

The giant manta ray was also fairly common, often passing leisurely beside the ship, the white bellies of the upturned remora (sucking fish) showing a startling white as they clung to the upper side of the ray's wings. If they fastened themselves to the underside of the ray it could scrape them off against a reef, and so to rid themselves of these parasites the rays (some of which grow to about fifteen feet across) flung themselves high into the air, turned belly up, and landed flat on their backs with a terrific splash. At night the crash of distant landings sounded like gunfire.

At dusk, probably exhausted by the heat of the day, young birds like terns fluttered round seeking rest on the decks, the mizzen boom, or the bowsprit. They seemed quite unafraid, and if it was necessary to change the jib during the night I first collected three or four roosting birds from the bowsprit, simply picking them up one by one and placing them in the shelter of the cockpit, where they stayed till dawn. One gusty evening a bird tried for half an hour to land on the plunging bowsprit, until just before darkness in desperation it flew straight in on to my shoulder as I sat at the helm. I put it on my lap inside my oilskins sheltered against the spray, and in the morning fed it from flying fish collected from the decks. It refused these until they were cut up and actually placed in its beak.

That bird stayed several days and became completely tame. On its head was a thick cluster of bloated ticks, an annoyance it could not remove itself. I took a pair of tweezers from my medical chest and picked the ticks off one by one. The heads of some were deeply embedded in the bird's skin and their extraction hurt; when this was so the bird held my finger firmly in its beak, before allowing me to extract the next.

Following my usual procedure of reading something of each country before arrival there I had bought a book on New Zealand, because I was to return there after being away from it most of my adult life. The book was written by a reputable socialist holding a high post under a socialist government and its object was to give the world a wide but general idea of why New Zealand was the best little country in the world.

My resentment was first aroused by the author's frequent generalisation 'we New Zealanders think,' often in direct opposition to what I as a New Zealander did think. It carried a blatant pride that New Zealand led the world towards the attainment of the perfect Welfare State, in which we did less work for higher wages than anyone else. And we were all terribly equal, a pleasant mediocracy where no man had more sense of responsibility than another, no one more brains than his neighbour, and where such refinements as the wearing of evening dress were discouraged because they aroused class-consciousness.

The undoubted scenic beauty of the country was expressed as a matter of personal pride, as if due to some creative effort on our part, with no sign of gratitude for the God-given gift it is. If I had read that book before being born, I would have chosen my birth-place in anywhere but New Zealand; so I spat on the last page, closed the book, and threw it over the side.

In these days too I read and studied the Bible, so that before entering those countries where I intended to study Hinduism and Buddhism I would first know something of Christianity. I had met several converts from Christianity to those religions (and Mohammedanism), and it had seemed to me that they had been enticed by the truths in their new religion without realising that those same truths already lay in Christianity. It seemed to me even then, with the little study I had done, that the original truths taught by all the big religions are identical, and that the rivalries lie mostly in superficial matters of dogma and ritual. It also seemed that all the big religions have deteriorated to varying degrees among the mass of their followers because the respective priesthoods have become more intent on maintaining their worldly prestige than on expounding the truths in a way that people can understand. In some cases understanding is deliberately discouraged.

Breaths of the first southerlies reached me and once an unexpected nor'wester blew for twenty hours. This was a beautiful day, running under the curves of the deep red spinnaker sweeping up to the masthead, and both the P&O liner *Chusan* and the troopship *Cheshire* passed close, the rails lined with passengers, and faint cries of *bon voyage* reached me. I burst into the old Army song,

> A troop-ship was leaving Bombay,
> Bound for Blighty's shore,
> Full of old soldiers and over-tired men,
> Bound for the land they adore.
> Da dit dit da dit dit da,
> The long and the short and the tall,
> You'll get no promotion,
> This side of the ocean,
> So cheer up my lads love 'em all.

All during this voyage the rotten mainsail gave a lot of trouble. I do not suppose the man who accepted my order and the payment before I left England will ever know what he cost me in time, hunger, thirst, anxiety and disappointment. There are few things more depressing than to lose the benefit of rare winds by having to lower sail to make a repair, to haul it up with a sigh of relief only to see the patch tearing away from the rotten canvas. So you down sail and begin all over again. My diary records one such event:

> Becalmed at dawn, spars swinging in the big swell left by the night's wind. While brewing up the morning cup I went to shake out a reef and found a long tear in the luff. Took 5½ hours to

mend—very hot and difficult to hang on as *Sheila* rolls unmercifully with sail down. U.S. cargo ship *Steel Designer* kindly left her course to ask if I needed help—very decent of them. [It was unusual for a ship to leave her course to ask if I was all right; of those who did 90 per cent were American.]

1400 hrs. No wind—I'm going slowly mad. It's taken me ten days to get from Lat. 19 to here in Lat. 17, average of 12 miles a day. Now sails are up to catch any breath, but the spars are crashing and banging from side to side until I'll go mad if it doesn't stop soon. Felt on the verge of heat-stroke after mending the sail and had to go below for a couple of hours. Took noon sight—have drifted back 30 miles since noon yesterday in spite of the night wind.

1800 hrs. The new patch in the luff has started to go—taken in, too tired to mend tonight, first thing in the morning.

As I crawled further south the prevailing southerlies reached me, averaging about force four and building up a short steep sea with decided stopping power. *Sheila* would just begin to get way on, run into a steep solid wall of water, check, pay off, gather way, and run into another. Her bottom too was getting dirty, which had a tremendously adverse effect to windward.

On the night of the 24th, a month out from Suez, I picked up the isolated light on the small rocky island of Jabal al Tair, and all next day beat south along a fifteen-mile chain of islands named Jubal Zubair. These low rocky mounds were all interconnected by coral reefs, shoals, rocks awash, and off-lying dangers.

The mainsail tore again at the luff and I spent the whole afternoon stitching the edges together and sewing a patch over all, *Sheila* pitching and tossing in the choppy sea while a strong current carried her close in towards the islands. A large aircraft carrier passed earlier on, looking very majestic and rather haughty in her freedom from torn sails.

At dusk I was about five miles off to the west of the southernmost island, and *Sheila* was sailing herself to the south under a gentle breeze. The lights of a steamer were just appearing over the southern horizon, and I estimated she would take about three-quarters of an hour to reach me. I was so exhausted by the short rations of food and water and the preceding days of heat that I was falling asleep at the helm. I had to sleep, and so setting the alarm for half an hour I went below intending to awake in time to avoid the steamer. It was against my rules to sleep on a shipping line and close to land, but there are times when sleep cannot be denied, and this is the greatest danger to sailing single-handed. I did not know it then, but only 100 miles away another single-hander was having the same trouble.

I awoke without the alarm and felt wonderfully refreshed, which was surprising because a body needing eight hours' sleep feels pain on being awakened after only half an hour. There was no ship in sight, *Sheila* was on her correct course, but in the moonlight the chain of islands instead of stretching away astern to the left, stretched away ahead to the right. Bearings from the islands, plus latitude from the North Star, put my position at five miles to the east of the northern-most island, or diagonally opposite to my position on going to sleep. I looked at my watch—3 a.m.! I had been dead to the world for eight hours.

Soon after I had gone to sleep the wind must have slowly changed, turning *Sheila* with it, and then sailed her the fifteen miles through that navigational nightmare of rocks, islands, shoals, reefs, and tide-rips. Sometime before 3 a.m. the wind must have returned slowly to its original direction, bringing *Sheila* back to her previous course to the south.

On long single-handed voyages there are times when in spite of all planning and care, over-fatigue takes its toll and then a man breaks his rules—and if you break rules you are looking for trouble. I broke my rule of not sleeping near land or shipping, and only sheer luck saved me from the consequences. Any single-hander who says, 'Alone I did it,' is talking the most arrant nonsense.

Another lucky break is recorded in my diary for 27 March:

Wind went S.W. early part of the night, then W., and by midnight was Nor'west. Was back on the shipping lane and sighted Jabal Zubar (the next block of Islands) by noon, and passed through the

Abu Ail Channel by 1600 hrs… full main and spinnaker all day, no time for food, terrible thirst; cooking up a big curry this evening to fortify me—I'll stay at the helm as long as this wind, lasts—only 100 miles to go to Bab-el-Mandeb.

Next day the wind dropped but then again went nor'-west, quite wrong according to the Pilot Book, and I sailed all night to pick up Perim Island light at four o'clock in the morning. At dawn I plunged through a tide rip off its southern end, right in the Straits of Bab-el-Mandeb ('The Gateway of Tears', so named for centuries past because of the ships lost among its currents and shoals), and almost decided to continue on to Aden, another 100 miles—the few buildings visible from the sea looked as if they had been shelled, but then a wisp of smoke showed and I reckoned to find some Arab fishermen who would at least give me some water, and perhaps bait with which to catch fish.

My small-scale chart was little help for entering Perim harbour, but the Pilot Book gave full details of landmarks, shoals, tides and reefs, after beginning '…Caution is necessary when entering Perim harbour…' That no longer frightened me because it applies to any harbour.

Once inside the entrance the harbour is spacious and mainly free from coral heads. I beat up to the northern end where the jetties supported a crowd of Arabs, and ran in close to ask if there was plenty of water; and just like people the world over they indicated that there was more than enough (without knowing how much I needed). The water was very clear so I took two more dummy runs to see for myself, and finally turning into wind, dropped sail and drifted up to the stone pier.

An attractive English girl miraculously appeared out of the throng. She had reddish-gold hair, was wearing those funny trousers women wear which are cut off too short; I even noticed that she had painted toenails, but I was too tired and thirsty to feel surprise. She said something about coming up to the Rest House for some coffee, and on the way said something more about her husband having gone off to rescue an American yacht, but not much of this registered. After the coffee she showed me a spare bed on which I flopped and slept dead to the world.

14

Another single-hander—Mike—a visit to Assab—Djibuti —arrive Aden

While I slept the husband came in from the sea, and awoke me in the late afternoon for a cup of tea. He was a huge, bearded person whose obsession to write had endangered the more practical application needed to live. He had done many things, from sculpturing in Paris, running a small farm in England, to his present enterprise of a fishing industry on Perim. He had an old launch and an Arab crew, caught sharks which were dried under the tropical sun and a cloud of flies, whereafter the bodies were sent to Africa and the fins to Chinese in Malaya.

He told me of his salvage venture. Petersen, an American single-hander on his yacht *Stornoway* (also designed by Albert Strange and somewhat similar to *Sheila*) had gone to sleep off Mocha (famous for its coffee) about forty miles up the Red Sea on his way to the north. He had stuck fast on a reef, been thrown into prison for a few days by the Yemenis, who had then allowed him to return to his looted ship. This information had come to the fisherman through his Arab crew, and he had immediately set off to salvage.

Petersen was in a very bad way when found, covered in rotten sores and seemingly in a hopeless situation. He signed a chit agreeing to pay £750 for being pulled off the reef, which he was, and taken to Assab. He later (I was told) got £500 from the States through the American Consul in Massawa, proceeded on his way north, and has by now arrived home after a complete circumnavigation of the world.

He was a quiet person, but seeing me with his expensive benefactor perhaps aroused suspicions of the interest I showed. In all ports I visited later where Petersen had called, people held him in high regard as very genuine and unassuming and a fine seaman.

The only other European members of the community were a young Greek with some vague interest in the fishing, and a young Irishman with a mad look in his eye. He aroused my curiosity because he was supposed to be walking round the world, and his presence on a small island hardly seemed conducive to the object. It had all begun because he had been too wild even for Ireland, and so an exasperated father had given him £200 and a first-class ticket to South Africa. There Mike had immediately made many brief friendships, whereafter he had bummed his way up Africa to the Gold Coast and then across the Sudan to arrive finally in Djibuti in French Somaliland. There, possessing only the shorts and shirt he stood up in and a battered attache case holding his manuscript, he took up residence in the public park.

A Somali taxi-driver had then befriended him, arranged his meals at a café and found him free accommodation in a brothel—brothels in the Red Sea area (so I was told) are much more free and easy than the relentless money-grabbing establishments in the West. The fisherman had met him in Djibuti while unloading a catch of fish, and had brought him back to Perim to work for his keep.

Next day we all went to Assab. It took six hours to cross the Straits, whereafter we swung northwards parallel with a long coastal reef. It was easy to see this from a distance of several hundred yards while the sun was behind us, but later with the afternoon sun beyond the reef the coral could not be seen until right on top of it. This is the first lesson to learn for coral navigation. We rounded a lighthouse on the Island of Fatma, wended a way between dark green mangrove islets, and arrived in Assab in the evening.

That night accommodation presented no problem. Mike and I went to the local dance, which for us ended about ten o'clock next morning. At that time we were sitting on the mud floor of a café in the suk (native quarter) owned by a blind Greek and his Arab wife. Whenever we lit a cigarette for ourselves, we lit one extra and pushed it between the lips of our blind host. A bottle of Anis (*Zib-zib, oozu, arrak, annisette*—it is drunk all through the Mediterranean and to the end of the Red Sea) stood empty between us, and hunger, too, closed our philosophical discussion.

The payment expected by the fisherman had not materialised and he refused to return to Perim until it did, so he arranged credit at an Italian café snuggled inside a burnt-out shack with holes in the roof and a verandah supported by luxuriant bougainvillea. The proprietor stated that he had to pay no tax for a café licence, as even in official eyes the place did not constitute a building. For one whole week we lived on nothing but fish and spaghetti for breakfast, lunch and dinner; we wandered aimlessly about the streets and the nearby beach, sat penniless in cafes, talked among ourselves, and time lay dead in our hands. I had a six weeks' beard, and was reduced to one pair of shorts and a shirt which after a few days were foul with dust and sweat. I had nothing with which to clean my shoes, and the laces broke. Perhaps it was not surprising that the British Commandant would not cash a cheque for me on funds awaiting me in Aden. Nor would he allow me to sleep in the empty Government Rest House, but benignly permitted me to go there once a day to use the bathroom and wash.

The fisherman slept in a local pension; Mike with his incredible charm slept on credit somewhere down the suk. The Arab crew and the unsold cargo of acrid fish made the launch unattractive, so for me a place to lay my head was a problem.

That morning I went to the Rest House to wash, and the wretched Dankali caretaker had received instructions to let me in. He grudgingly unlocked the bathroom door, making his authority very much felt to one of the master-race rejected by his own kind, instructed me not to take too long and to call him to lock up when I had finished.

As soon as he had gone I investigated a door leading into a room off the bathroom; in it was a table and chairs stacked in one corner and a pile of mattresses against the wall. I prodded them with a chair and heard the dry rustling of disturbed scorpions. The window in the outside wall was closed by a heavy shutter, with bolts at the top and bottom. I lifted the bottom bolt on to its catch to leave it unlocked, but hardly noticeable to a casual inspection, and the top bolt I eased on to the very edge of its catch, so that a shake would let it drop into the unlocked position. Then I returned to the bathroom, locked the intervening door, washed, called the caretaker and departed.

After dark that night, and others, I waited in the shadows until the caretaker had his evening meal, did his rounds, and crept into bed with his hideous wife. Then I went to the shutter in question, gave it a gentle shake and heard the bolt fall; I then climbed inside, shook out a mattress, and slept until four in the morning before the caretaker became conscious. Setting the bolts in the reverse order I climbed outside, shook the shutter, and the bottom bolt fell into the locked position. This was readjusted when I went for a wash later in the morning. And so the caretaker slept happily in the belief of duty fulfilled, and I slept happily because it was the best place I could get.

Mike wanted to get to India by sea, rather than by trudging hundreds of miles over waterless wastes. He had already been as far as Aden on an Arab dhow, but the authorities had refused him permission to land and he had had to return. I admired his venture and offered to take him on as crew as far as Bombay, knowing that such individualists sometimes have one theory for expression in intellectual discourse, and another for practical application. I felt that I could deal with this contingency if it arose, but I hadn't then seen Mike in action. First we sailed to Djibuti where he had left his manuscript.

Djibuti is about 80 miles from Perim, but owing to light winds we did not sight the lighthouse on its guardian coral reefs until dusk on the second evening. Although I had no large-scale chart, the Pilot Book gave its usual detailed instructions for entry, and the reefs could be avoided in the darkness by bearings from the various lights. I decided to go in that night rather than wait for dawn.

The night fell softly dark and *Sheila* sailed herself gently across a warm breeze. Mike and I were talking in the cockpit when we noticed a light far out to the south-east. As we watched it grew more vivid and was seen to be sweeping towards us; it seemed like the beam of a very powerful lighthouse, pivoted in the south, and sweeping from one horizon to the other—but under the water. It rapidly came closer, relentless and inexplicable, until it lit up the sails with a greenish light quite bright enough to read by. I watched the defined beam as it passed under *Sheila*, throwing the dark shadow of her hull momentarily over the sails, and then it fled to the western horizon.

It left us speechless, but another great beam appeared in the east, swung towards us, underneath us, and silently fled into the western darkness. This happened about five times, always the same, at the same regular intervals, in complete silence and with no change in the wind or sea. Mike, a devout atheist, uttered a shaking blasphemy, and admitted that one more beam would have put him on his knees.

Early next morning I sent Mike ashore in the dinghy to get his manuscript and some food while I got *Sheila* off a mudbank on which we had stuck the night before. He spent the time ashore to good purpose, but returned with only a tiny loaf of bread for me. Later that day other complications arose, and I began to think that Mike and I must part for our mutual peace of mind. I had no intention of changing my way of things, and had no desire nor right to change his—which passed unnoticed in a place like Djibuti but would be embarrassing in Aden and Bombay. It surprised me very much when Mike insisted on accompanying me to the police HQ where I had to report my arrival.

The Gendarme inspected my passport, gave me permission to stay for four days, and then turned to Mike.

'I'm his crew and leaving with him,' said Mike.

I would have preferred another time and place to decide our personal affairs and Mike had forced the issue hoping I would evade it, but once I accepted it before the Gendarme it would have been even more difficult to refute later. I would also be held responsible for Mike's doings in Djibuti, and he was capable of tearing the town apart. So I said to the Gendarme, 'Non Monsieur, il n'ira pas avec moi.' I was allowed to go but Mike was held to answer questions. It seemed a ruthless thing to leave a man penniless in a foreign port, but Mike was quite used to looking after himself and would have quite casually involved me in much more difficult complications. As his skipper local authorities would hold me responsible for his conduct, and I had reason to assume that Mike was not amenable to discipline.

Later that day he swam out to *Sheila*, streaming with blood from cuts received in falling down the rocks, and tried to make me change my mind. I knew we would both regret it if I did. There was a scene in the small cabin, but in the evening Mike appeared again with a box of stores.

'No strings attached,' he said. 'I'm sorry. You're quite right. This stuff will get you to Aden.' A strange person, very likable, but possessed of a devil he will find it hard to banish. I often wondered if the Church he had forsaken can.

Djibuti was a place such as I had never seen before. There were many Europeans of many different races who seemed to spend most of the late mornings, early afternoons and every evening in the numerous cafés, apparently living on the profits of a small counter on the street from which a small boy sold fountain pens and watches for them. It is not done to ask questions in such a place but all talked to me about the others, giving the impression that they had not so much a particular reason for being in Djibuti as a better reason for not being somewhere else.

A troopship from Indo-China called and debouched its Foreign Legion contents for four hours into the suk. By two o'clock every day the heat became unbearable and the town slept. The small bands of Ethiopian and Somali prostitutes, bulbous, brazen girls, had swept the town searching for anyone with nowhere to siest; the café tables were deserted, with the chairs upturned on top of them; Arabs slept on the pavements, on the backs of parked lorries, in silent cars. And the clouds of flies sought undisturbed satisfaction amongst the rubbish in the gutters, the sticky stain of spilt wine on the tables, the littered pavements, and the sleeping bodies.

I sailed out of Djibuti at dawn on 18 April with a fleet of dhows, idling through the calms to arrive off Aden three mornings later. The Pilot Book gave instructions that all shipping must anchor off Steamer Point for health clearance, and there I saw about twelve ships with the yellow 'Q' flag flying, showing that they were yet unattended. I hate queueing up for anything, and so sailed through the lot under full main and the deep red spinnaker to find an anchorage behind the boom defence pier in Aden harbour; within ten minutes my 'Q' flag brought the Port Health Officer on his way out to the steamers.

Many, many times have I criticised the dull convention of British Colonial administration; but after all I had lately seen elsewhere it brought a relief like the sanctity of a home.

15

A hard decision—I love the Army—departure

Aden is, I suppose, one of the most barren areas of the world, nothing but sand and bare ruthless rock rising sheer to jagged pinnacles of nearly 2000 feet. My time there was initially dominated by two pressing problems. The first was to have *Sheila* slipped, because she had been in the water six months since her last anti-fouling in Algiers. And the second was to decide how to get to India, because I had missed the fair-weather monsoon.

The tides in Aden did not have sufficient rise and fall to let me beach *Sheila* to clean her off. There were no slip-ways, but I discovered an Indian firm which owned a floating dock, and they kindly agreed to accommodate *Sheila* behind a small tug which was to be raised for survey.

The second problem, that of getting to India, was more difficult to solve. In that part of the world, as in most other places, the year is divided into four seasons: the fair-weather or north-east monsoon runs from about January to March; from April to June is a transition period of flat calms, variable unreliable winds, and tropical cyclones. These lead into the south-west monsoon from July to September, which sweeps unimpeded across the thousands of miles of open Indian Ocean averaging near-gale force, and building up huge destructive seas. The last season, from October to December, is another transition period with the same conditions as the first.

I had decided after my intensive reading in England that providing I looked after *Sheila* and maintained her properly, the sea in any mood would give me a fair chance—except in a cyclone (or hurricane, typhoon, they are all the same thing under different local names). Too many big ships have been overpowered in these super-storms, and although some small vessels have survived them the odds are heavily against it. I felt that I would be looked after during this voyage, but to cross a stretch of ocean at the height of the hurricane season would be tempting Providence.

So I ruled out sailing in the transition period, which then left only one alternative to foregoing the trip altogether—to sail in the Southwest Monsoon. I concentrated on finding out all I could about it, to extract the greatest advantages and disadvantages, and thereafter planning how best to use the former and how to avoid, or at least know what precautions to take against, the latter.

The Pilot Book told me that all the smaller ports on the west Indian coast were closed by the monsoon, and that coastal traffic ceased. Its description ended simply, 'Small vessels do not put to sea.'

The greatest advantage of sailing in the south-west monsoon was an assured hard wind on the starboard quarter, meaning a fast passage. The first disadvantage was the consistent heavy weather to be expected; in the past a week's gale had left me exhausted, but on this crossing of approximately 2000 miles I could expect those conditions for all of it, a period of perhaps four to five weeks. Another disadvantage of heavy weather is that high-flying spray keeps all ropes and canvas wet, when any chafe causes much more damage to the rubbing parts than when they are dry. To neutralise this danger I renewed such of *Sheila*'s running rigging as I could afford.

The danger of exhaustion I planned to eliminate by not trying to make a fast passage. It is always tempting to sail long hours with a fair wind, but I determined to take plenty of time off for adequate sleep and in which to prepare good, hot, well-cooked meals. As far as *Sheila* was concerned with the rough weather, I had little doubt of her ability to deal with it.

The other danger I extracted was the landfall on the Indian coast, but that was something I could only deal with when it arose. Yet in spite of all this planning—and for the two and a half months I waited in Aden these problems were never far from mind—in my inexperience I completely overlooked the greatest hazard of all. That too will be told.

It is easy to write simply of those problems now, but at the time I was torn first one way then the other. From all advice the south-west monsoon meant certain death and failure, and I stood alone in my decision to attempt it. Some kind local people formed a body and asked the harbourmaster to prevent my sailing, but he had no authority over private vessels. Such control must never be—what if Columbus had not been allowed to sail over the edge of the world? The importance of that venture was not the discovery of America by Europe (a statement with which the few remaining Red Indians will probably agree), but in the fact that a man was allowed to fulfil his belief. If that is stopped, all is stopped.

In moments of depression I seriously considered deserting *Sheila* and applying for a passage home as a Distressed British Subject. It is sometimes very hard to know where the line lies between healthy determination and stubborn stupidity, and this is when loneliness is known, deep and ravaging, when you doubt yourself. And in those times I longed for security, for the easy, irritating, monotonous routine which usually brings the reward of a steady income, and the release from responsibility of making vital decisions alone.

One event at this period brought great joy, when the *Empire Trooper* steamed into Aden bringing two friends in my regiment, two such friends as are a blessing to any man. There were also other officers with their wives, of other Gurkha regiments, and that evening as we sat on deck talking the days I had forsaken came back so easily, bringing with their joy a greater sadness for their loss.

At 10.30 p.m. a voice of authority came over the ship's loud hailer system, 'Clear the decks please.' And soon the Orderly Officer, a mere Lieutenant, did his rounds to hustle those loath to obey. I and a number of adult people hastily finished our drinks and disappeared to the stifling cabins like clever sheep.

I love the Army, I miss it terribly and always will, but not this. I returned to *Sheila*, sad, but content with the decision I had made in 1949.

Most of my time in Aden, apart from that spent preparing *Sheila*, was passed at the Gold Mohur Swimming Club. This is merely a collection of changing rooms and a portion of beach wired in against sharks, with a diving platform in an outer corner of the net.

And then one day my new mainsail arrived from England, nearly a year after it had been ordered and paid for; but then followed pleasant days in the harbour stretching the new canvas, usually accompanied by friends and a few bottles of beer on ice. Sometimes we sailed round to Gold Mohur and spent late hours at the swimming club.

The last days as always were a frantic rush, mostly saying goodbye to people—some had dined me and taken me out on parties, some had said, '...you can't live on board. We have a spare bed, come and go as you please, don't bother about being in time for meals, just do as you like.' Others had helped me with the myriad details of *Sheila*'s administration; a girl had spent hours patching my torn spinnaker, a pilot had got me charts from ships in transit (the average cost of a chart is about 7/6d). All did their kindly best to deter me from sailing, and some even offered me jobs so that I could afford to stay for better weather. But by that time my one determination was to get to India with that monsoon, whatever its fury might impose; it became a certainty, with some apprehension and the acceptance that it was going to be sheer hell.

I rowed out to *Sheila*, got the dinghy on board, hauled up short on the anchor, raised sail and let *Sheila* break the anchor out. She lay hove to while I lashed it on the foredeck and coiled the halyards, and then I sailed once past the small crowd standing on Post Office Pier, and headed out to sea. I felt as if I were going in to an attack, but it was going to last longer than any I'd known in war; nor did I know that before reaching Bombay *Sheila* was to meet three occasions on any one of which the voyage might have ended. And the first was only a few hours away.

16

*Nearly wrecked—the need of a writer—nearly wrecked—
Mukalla—departure*

And so I sailed out of Aden on Sunday 17 June, 1951.

The engine was still out of action, which worried me little. It saved the expense of fuel and stowage space, which for that bulky commodity was always a problem. I did not like to have petrol below decks because of the risk of fire, and so one four-gallon tin was usually carried in a corner of the cockpit. In the other corner was another four-gallon tin of kerosene for the primus. Even these two tins were a nuisance, as most of the working of the ship is done from the cockpit; more a mental nuisance, like trying to write on an overcrowded desk, than anything else.

Four other four-gallon tins were on the side-decks in pairs lashed to the main port and starboard stays. They were in the way there when I went forward to change jibs, but less so than anywhere else. So apart from three gallons in the tank, the usual petrol stocks were 20 gallons or about 100 miles' cruising range. This is nothing in a 2000-mile voyage and was used mainly to charge the batteries for the electric light, but even this was not important. Everything on *Sheila* from a primus pricker to a large spanner had its one and only place of stowage, and I could always find what I wanted in a moment, sometimes going for weeks without using a light at all. The compass light was the most important, but with clear skies at night I steered by the stars.

Most think of an engine as the driving power required to take a ship away from a stormy lee-shore, but very few auxiliary engines have the power to drive a yacht, with the surprising resistance of its masts and rigging, against a heavy wind and sea. If there is wind a good yacht does not need an engine, and I would sooner be on some lee shore in *Sheila* under storm canvas than in the usual motor-boat with its not infallible engine. It is when there is no wind that a yacht needs an engine, and courses are planned taking this into consideration, but the first two of the three dangerous incidents of this passage will illustrate clearly what is meant.

The light wind took me along the coast towards Gold Mohur on my way to the east, until shortly it died altogether. There was a strong on-shore current and a big swell lazed in from the open sea, but I was in no danger; it was a sandy bottom and I intended to anchor in about three fathoms if wind did not come. A car had been following along a coastal road, and it halted on the beach towards which *Sheila* then drifted with lifeless sails.

A girl waded into the surf as I went forward to let go the anchor in about twenty feet of water, and still about sixty yards off. There was a slight wind, but not enough to tack *Sheila* against it over the rollers which shouldered her towards the beach.

I heard a shout from the girl, she waved, plunged into the surf, and to my horror started swimming strongly out to *Sheila*. Her male escort, chivalry overcoming sanity, followed. This was madness in those waters notorious for sharks—I had grown fond of that girl but I could have beaten her then with pleasure. She would probably have got to *Sheila* all right, but this was a situation where I had to do what I could while I could still do something, because if a hungry shark did arrive I would then be powerless to do anything. The only action possible was to lessen the time the girl would be in the water.

Sheila turned slowly with the light wind and I took her inshore to meet the swimming figures. The bottom was already clearly visible, and I could only hope to get these people on board with still enough water in which to anchor. (A smooth, sandy beach looks innocent enough but a ship laid over on her bilge, pounded by a heavy surf, can be severely damaged in a very short time and I had no money for salvage nor for subsequent repairs.)

The two struggled on board, elated by the risk, only to be nonplussed and hurt by my decidedly cool reception. They had no inkling of the danger in which they had put me, and it was too late to explain—as *Sheila* sank on the swell I felt the tremor through the hull as her keel touched the sandy bottom. Only the fastest work could save her, so I first hurriedly disposed of my guests over the side (we were then in only 5 feet of water). The man offered to help but I can work faster alone than with an untrained companion and spectators get in the way.

As the pair splashed over the side I clawed down the sails, unlashed the dinghy and flung it overboard, made fast a rope to *Sheila*'s bow and started to row, to tow *Sheila* into deeper water. As she sank on each lazy swell I could see her masts and rigging quiver from the shock as her keel hit the bottom.

Slowly her bow came round, but there was no perceptible way ahead and when the swell lifted her bow high in the air, pushing her back towards the shore, the dinghy was dragged helter-skelter after in spite of all my labours. Sweat streamed into my eyes, the muscles in my forearms knotted into a fiery ache, and the only thing which kept me going, beyond hope itself, was simply that this was the only thing I could do. Once *Sheila* got some way on ahead it would be easier to maintain the momentum.

An age later I anchored in three fathoms, and with dusk came a breeze which carried me round the corner to Gold Mohur. In the darkness the three of us spread a rug on the sand and talked far into the night. I was very tempted to stay just one more day, but if departure was to be hard that night it would be that much harder after another day. There had been beauty in the green surf breaking on the innocent sand, and the end of my voyage also; so, too, in this other.

Next morning I awoke with a raging temperature and a terribly upset stomach. Perhaps it was a tropical bug from some of the stinking water drunk in Burma, perhaps it was the result of an emotional crisis, which always upsets my stomach. It was madness to begin a hard, long voyage in that condition, whatever its cause, but at sea I had come to seek a guidance beyond logic on which to make decisions; it can be infallible but the conflict between an unformed faith and an obvious logic can be dangerous to your life if you go against the one, and your sanity if you go against the other. Was this raging temperature a sign that I should return to Aden as part of me so strongly desired, or was it a test to be overcome to prove and strengthen my belief? I shall never know, because as soon as one alternative is followed any answer from the other is denied, and so no sign is given.

I got in the anchor and was violently sick while hauling up sail but determined to go east, believing that to look back is disastrous, more so than the danger of physical weakness at sea and all its consequences.

The Gulf of Aden is only on the fringe of the monsoon and hardly affected by it. The heat was appalling and the long hours of calm, or running under the spinnaker at almost the same speed as the light wind, sometimes brought me close to heat-stroke. Even at night my naked body streamed with sweat while the dark hours were spent on the lookout for shipping, but after nightly prayers for wind each dawn stole into a leaden calm.

Yet the sea brought peace; I had made my choice for better or for worse, and there was no turning back. For the first days at sea I was thus always on edge until the painful transition from one extreme way of life to the other was completed, when the more obscure values of solitude at sea banished the more obvious values of security on land.

I had been reading Stephen Spender's *World Within World*—the need for a writer to be independent, to be himself unshaped by convention or other people; and yet the need for a writer to experience people to gain more understanding. It is in that very experience, the acceptance of attention, of love even, and so of obligation, that the greatest threat to his freedom lies. If he has great conscience about the hurts he gives to others in refusing to be chained, he will be chained; if he has little conscience he destroys an essential part of himself. He can only feel the pain he gives to others deeply and bear it with them, without the solace of their knowledge that he does so.

The fourth day out is recorded in my diary and is typical of those at that time:

Been a bastard of a day, heat beyond belief, no wind except a faint puff now and again from the S.W. Also a big swell from the S.W., so perhaps I'm on the fringe of the Monsoon at last.

The medical front is not so good. The rope burn on my knee (from sliding down the mast too fast in shorts) has gone black, and the scab seems to be eating a deep pit into the flesh—I pulled it off and filled it with disinfectant. My feet have swollen and cracked to raw meat between the toes—I don't know what to do with them. My tummy is still dreadful—I drank a can of tomato juice and within five minutes it was over the side, still as tomato juice. I might just as well have opened the tin and poured it into the sea direct.

I hope it's not too much of a test getting to Bombay—I'm weak now from this bloody stomach, so what a gale will do to me I can't imagine—that's just what's coming brother, weeks of it, and so you'll soon know! [I often talked to myself in this way.]

After a week out I definitely decided to call at Mukalla, a small town on the south Arabian coast about 200 miles east of Aden. It was wise to top up the water tanks as consumption had been heavy; I wanted the experience of being a lone white man in an eastern state under its own law and potentate; and I wanted to call where the world-famous sailor Robinson had called.

I studied the Pilot Book and chart for information about the approach and anchorage. The Pilot Book told me that there was an onshore current running up to nearly three knots (a fair walking pace); and the chart showed a coral reef running parallel with the coast and about a mile offshore. It rose sheer from depths of eighty fathoms, far too deep in which to anchor, so if the wind failed the current would sweep *Sheila* on to that reef.

The white buildings of Mukalla showed above the horizon about noon, looking like a thin line of surf at the foot of the hills. I kept my course parallel with and about fifteen miles off the coast; it was too risky to cut the corner and head straight for Mukalla in case the wind dropped, to leave me close inshore at the mercy of the current. At three in the afternoon, when I was about ten miles short of Mukalla, the wind did drop.

A huge oily swell rolled in from the south (a full hurricane had been raging 200 miles away off Socotra); the air was dead and remained so. I did not worry overmuch as the current would take at least six hours to carry me on to the reef, and surely by then the wind would come?

Bearings taken at five in the evening confirmed the rate of drift as 2½ knots. There was still no sign of wind, and a continuous angry growl hung in the air—the heavy swell breaking on the reef. I was in a situation of very great danger, because if wind did not come within three hours, it looked like certain shipwreck.

Night came quickly as always in the tropics, and the soft darkness lay thickly over *Sheila* as the current carried her closer to the reef, from which there came a steady roar. There was still no breath of wind, but the huge swell came in lazily, monotonously, relentlessly. I started to take soundings but without avail, until at 9 p.m. the lead found bottom at thirty fathoms, a lucky out-crop not shown on the chart. I hauled in smartly and sounded again and continued so, each sounding showing a decrease of one fathom, sometimes more. The bottom was shelving steeply and at eleven fathoms I dared not delay longer. The heavy crash of each individual breaker could then be heard, and the darkness glowed as with lightning below the horizon, the phosphorescence created by millions of tiny creatures startled into light when their vehicles crashed to destruction.

Sheila checked as she came to the end of the rope, and I returned to the cockpit to take another sounding. It showed 6 fathoms and I left the line down to indicate whether the anchor was holding. Apparently it was but I was still not safe—as *Sheila*'s bow lifted feet in the air with each swell, straining hard against the current as well as the lift, she could easily pull the anchor free and there would be no second chance. I estimated the reef was about forty yards astern, which at the rate of drift was only about thirty seconds in time; and the regular sawing strain on the rope could cut it on a jagged coral edge.

My second anchor was a CQR designed for holding in sand or mud, and quite unsuitable for a hard bottom. This was my only reserve and for this reason I had left the chain shackled to it. The weight of chain allowed it some hope of holding, but with rope in that depth of water it would have been useless.

It was no use simply throwing this second anchor over the side, because if the first one did give, by the time *Sheila* took up the slack on the second she would be on the reef. It had to be taken out as far ahead of *Sheila* as possible.

By this time, due to the preceding week of dysentery and the fantastic heat, I was weak and very tired. The thought of laying out the second anchor for such use as it might be was far from pleasant. It was a severe temptation to creep into my bunk saying, 'I believe the anchor will hold'; but this I knew would have been self-delusion and merely shirking the issue. Why should belief have power on one anchor when I had two?

I unshipped the dinghy, clambered in, and pulled it forward to *Sheila*'s bow. One second *Sheila*'s gunwale was level with the dinghy's and the next reared high into the darkness and beyond my reach. The descending bowsprit twice caught me across the shoulder and knocked me into the bottom of the dinghy, nearly upsetting it. Yet somehow I was able to lift the heavy anchor from the deck into the dinghy, and draw the thirty fathoms of chain after it.

It is not possible for a man alone to lay out an anchor on chain efficiently in a dinghy. As soon as he has rowed a few yards out the weight of the intervening chain draws away the slack with increasing speed, and tends to pull the dinghy back to the mother ship. The chain flew over the side of the frail dinghy, splintering the thin planking, so I dropped the oars and grabbed the chain with my bare hand. The escaping links tore the skin from my palm and the slippery blood made holding an impossibility, so taking the oars I rowed with all my might into the darkness and flung the anchor over when I reached the limit allowed.

Back on board *Sheila* I took another sounding. She had pulled back into 5 fathoms but had taken up an even strain on both anchors, the chain and the rope being two phosphorescent lines reaching far down into the blackness of the sea. There was nothing more I could do.

I lit the lamp in the cabin, bound up my hand, brewed up a mug of coffee, and smoked a cigarette. If the anchors did not hold, *Sheila* would be smashed to pieces on the reef, when it would be too late to escape in the dinghy because no dinghy could survive there. After being torn to mincemeat on the coral I doubted my chances of swimming the mile to the shore through inky-black, shark-infested waters, while bleeding like a stuck pig. The safest bet was to sit in the dinghy beside *Sheila* and, if the anchors dragged, keep outside the breakers and row the ten miles to Mukalla.

In moments like this, when alone and untrammelled with ordinary logic, you feel that only belief can save you and your ship. This belief does not come because you want that which it will give; perhaps it comes only after you have done everything humanly possible to ensure by your own efforts that which you now need from a higher power.

So having done all I could I felt strangely unconcerned by the fact that any moment, only thirty seconds away, might come a violent and painful death. To sit in the dinghy was rather like saying, 'I believe—but with certain precautions in case it doesn't work.' And so I lay down on my bunk exhausted and fell into a deep sleep.

Wind came soon after dawn, which showed the angry reef indeed only forty yards astern, and I wondered gratefully that I had spent the night sleeping deeply. I got on sail, took in both anchors and sailed for Mukalla.

After reporting to officials (who were mostly Pakistanis and spoke Hindustani), I wandered about the beaten earth streets of the town. The tea-houses were deserted, and when I entered one to ask for tea the man shook his head violently, saying, *Musht tamam* ('No good'). I took this to be colour or racial discrimination a new way round, so returned to the street once more and lit a cigarette. Immediately an angry local rushed upon me, snatched the cigarette from my lips and stamped it angrily into the earth. He also among other things shouted *Musht tamam*—and then I caught the word *Ramadan*. This is the Mohammedan month of fasting, when from sunrise to sunset they neither eat nor drink nor apparently smoke, and expected me to follow suit. Any man found breaking the fast before sunset was liable to be stoned to death by the crowd. This also explained the numerous women in the streets, all heavily veiled but unescorted, which is unusual in a strict Mohammedan community—women were safe from any man during that month between sunrise and sunset, as this too was part of the fast imposed. The fast was relaxed between sunset and sunrise.

In Mukalla of all places I met a fellow New Zealander, an official of some sort in the Administration. He was very hospitable and it was in his house that I met Fatima and Farid, two Arab children of about five and six years of age. Several times I found them giving me that open, level stare which children give a stranger, and I waited patiently for acceptance. By evening they were showing me their toys, leaning against me with the affection children so generously give when trust is known, and behaving exactly like children anywhere else in the world.

On the third day after arriving I was ready to sail, and as the wind and sea had freshened into the open anchorage from the south-west it was no longer safe to stay. I bade farewell to my New Zealand host, to His Highness the Sultan of Mukalla and Socotra who had kindly received me, and to my small friends Fatima and Farid. *Sheila* beat out against the rising wind under two reefs, and turned her bow into the open sea towards Bombay. On the end of the promontory of Ras Kodar lay the bones of a huge dhow, wrecked there four days previously.

17

*The south-west monsoon—exhaustion—a friend—
dangerous landfall—arrive Bombay*

The first three days out of Mukalla were generally of moderate winds broken by powerful gusts which came at any hour of the day or night, but on the fourth day out the wind rose to a steady gale force, and I knew that I had then entered the area of the established monsoon, where it blows unceasing for three months over the expanse of the Indian Ocean and Arabian Sea. The great lazy swell became more vicious and for the next three weeks *Sheila*'s decks were never dry. Each day and night were the same as the one before and after, except for slight variations in the wind, which moderated to average near gale force, always enough to tease the top of a huge roller into a breaking crest, which flooded seething over the after-deck, climbed the coaming and filled the cockpit. The hatches were closed night and day, whether I was above or below decks.

In these days the greatest hazard of the voyage developed, one which in my inexperience I had completely overlooked—salt water boils. The continual soaking in salt water affected my skin, until the flesh of my seat and thighs particularly became sodden and puffy and was completely covered in scores of small boils. It was too rough to stand and steer efficiently, and sitting became an agony I cannot describe. Many may know the pain of sitting briefly on one boil; I had to sit for hours on about a hundred.

At times when this became unendurable I went below, stripped off all clothing to escape its touch, and stood naked in the cabin. On one occasion I thought that methylated spirits might dry and harden the skin, so poured some around my naked waist. The raw spirit stung like fire in the open boils, and trickled into various crannies and orifices of tender skin not designed for contact with such fiery liquids. I never tried it again.

Then one splice holding a main lower shroud to the hounds partially pulled out (this splice had been done by a qualified bosun in Ismailia), which in that weather left the mast in very great danger until the shroud was replaced. I took out a length of heavy flexible wire kept for just such an emergency, with an eye-splice already in one end. Owing to the time I would have to be aloft I decided to use the bosun's chair, to give me a platform from which I could work with least difficulty.

I had to grip the mast tightly with my legs to avoid being flung far out by the sway of the mast, only to come crashing back against it with the opposite swing. One arm also encircled the mast and I don't know how I pulled myself aloft, but once there the wire was looped round the mast and drawn through the eye-splice, and fastened with seizing wire to form a tight loop over the hounds. I dropped the free end to the deck and lowered away. On deck I slackened the rigging screw, fastened the wire with bulldog grips, and made taut with the rigging screw. Then I went below to strip off my clothing and rest. The inside of my thighs was a mass of open slimy pus, the heads torn from the many boils, and the soft white skin had pulled away in chunks to leave patches of raw flesh.

It was imperative to go aloft again, something I would have given almost anything to avoid, and this was an occasion when I was glad no one else was there. If he had gone aloft and fallen to drown in the seas I would never have forgiven myself; if I had gone aloft I would have resented him, thinking, 'Why is he not a man I can rely on implicitly?' What human can be relied on implicitly when the tiniest slip means disaster? And so it is better to do the job alone so that neither remorse nor resentment can enter into it.

If one splice had gone it was likely that its opposite number was also weak, if not from the beginning, then with the extra strain it had had to bear. Even heavier strains would be imposed on it by the more vicious seas to be met inshore when I made my landfall, and if it went then, it would be both impossible and too late to repair. A broken mast on a stormy lee shore means certain shipwreck, because anchors would not hold in the seas to be expected. This job had to be done, and done at once before excuses grew.

This time I hauled myself up on the main halyards and sat on the cross-trees astride the mast. I could only work with one hand and the nearest foot in moments of comparative equilibrium. The violent jerk at the end of each swing of the mast, as *Sheila* plunged over a crest or rose from a trough, demanded both hands and all else to stay put, while not dropping the wire nor losing that amount of work completed; yet the longer I stayed aloft the less strength remained to get me down again.

I crawled off the cross-trees, gripped the mast with my legs, and before starting down I concentrated, once again pinpointing my mind into my hands. I forgot the sea and the ship, the pain in my legs, everything but my hands, because they needed a strength my body no longer had. The smallest weakness and the clutching fingers would have been torn apart. Hands, hands, hands. Even now, as I write this story so long after, I can still remember the feel of hard, wet spiral strands of rope beneath them.

The job was well finished, the danger past, and peace as always came with having done that which so obviously had to be done. I went below to the cabin, my will relaxed but pain returned, and I sobbed with the agony of my diseased, bleeding legs. Leaving *Sheila* hove to I made no attempt to sail more that day.

The hours dragged past, sheets of spray flew over *Sheila*, and a black depression told that the end of endurance was near.

You cannot give up at sea by walking off the ship as you walk away from a job, but a mental depression is more dangerous in that it tolerates carelessness. If at that time the mast had again become endangered I doubt if I would have given a damn where it went. As darkness fell the mood increased, eating into me; it had to be broken and the only way to break it was by work, so I drove myself to the helm with orders to stay there till dawn, and make up the time lost during the day.

By 4 a.m. the spray had long since found its way into my oilskins, soaking all clothing to the skin; my back, arms and shoulders were aching with the effort of steering in the big quartering seas; and the pain of sitting I could no longer endure in spite of orders to stay till dawn. I gave up, beaten, and leaving *Sheila* hove to I crept to my bunk in the dark cabin, to fall asleep not caring whether I saw India or not.

'Get up. It's your turn at the helm.' It was the same voice that had called me to avoid the approaching steamer.

The firm tones awoke me, and as I lay in the bunk I could see through the open hatch (I'd forgotten to close it) to the helm. It was only just after dawn. At the helm was a fair-sized man; he showed no signs of fatigue nor worry about the weather. I told myself I was dreaming, yet his grey oilskins impressed me—all others I had seen to that date had been black or yellow. I felt wonderfully refreshed, which was noticeable at the time and surprised me, because I really needed far more sleep than the two hours I had been allowed. I got up, put on my oilskins, and wondering how on earth the man had got on board I went to relieve him at the helm. There I checked the compass (*Sheila* was hove to), altered sail, and put her on course. Somehow my friend had passed me on the narrow companion-way and was then in the galley, bending over the primus as if about to get breakfast.

'Where do you keep the eggs?' he asked. (Those were the exact words I heard and will always remember.)

I laughed and replied aloud, 'You know bloody well we haven't got any eggs.' My exact words too.

'This is very strange,' I thought to myself. 'No one else can be here, but he is here. I am awake and feel wonderful. Who is he?'

I pulled myself together and studied this man's face deliberately and intently as he continued to bend over the primus. He had a ruddy complexion, and the fringe of hair sticking out from the upturned brim of his sou'wester was sandy coloured. His nose was long and finely moulded, firmly defined lips, and a strong jaw; his most striking feature was his eyes, which were very blue.

He moved into the saloon where I could no longer see his face, but from the movements of his body I could see he was taking off his oilskins. It is where I always took off mine, as there is full head-room and a hand grip on either side.

This had to be solved, so I hove to and went below. The cabin was empty; feeling foolish but knowing I had seen him I looked in the fo'c'sle, the only other place on the ship he could have been. It was empty.

I felt a little sad, but his cheerful vitality remained. I felt strong and confident, glorying in the sea and *Sheila* as she surged ahead of the wind as joyfully confident as myself. I cannot remember now when the pain of my legs returned, but it must have worried me again.

There is no entry in my log or diary of this strange event, perhaps because I felt that to 'publicise' it would be a kind of sacrilege, like breaking a confidence; perhaps because at that time I was afraid of inviting ridicule—not of myself, but of it.

Each day the pale clear sky lay above the wide ocean of flying spray, and I waited for the first clouds as the warm moist air neared land. The first scattered cumulus appeared as tiny round puffs of cotton-wool, which each day grew and massed, soon towering into high white columns riding the wind, joining *Sheila* and me in our journey to the Indian continent.

Came the first rain and on 11 July, two weeks after leaving Mukalla, I took what were to be the last sights of the voyage as the sun showed briefly through gaps in the massing clouds. From this fix I had to find my way over 500 miles to a stormy lee shore and locate Bombay. Rain squalls increased until the torrents from above were almost unceasing, often reducing visibility to a matter of only a hundred yards or so. Within two days, under this deluge of fresh pure water, my boils disappeared and the flesh became firm and healed.

500 miles is a long way on DR only, in a sailing ship. If I steered for Bombay and missed it by only two miles in that visibility I would not know which way to turn along the coast to find it, whether to turn north or south. So I set a course to find the coast sixty miles south of it, to be sure of being south, and so when land appeared I would know that

Bombay lay to the north and the wind would be with me.

A week later I awoke at four in the morning to a different feel to the waves which told me land was near. On deck it was pouring with rain and as black as ink, but I took a sounding and finding no bottom lowered all sail except the jib, and let *Sheila* drift landwards while I got more sleep. It is wise to build up a reserve of this before reaching such a coast, where long hours may be demanded.

There was no sight of the coast at dawn but I found the bottom at seventeen fathoms, which put me roughly seventeen miles from land, as the sea-bed shelves on that part of the coast at about a fathom a mile offshore. I put on sail and continued towards the unseen coast, and later that day sighted land between rain squalls.

By evening I could distinguish low jungle-covered hills rising into the cloud behind; ahead lay a long headland surmounted by a fort. This landmark was little use, as on that coast almost every strategic headland is surmounted by a fort. These are massive stone buildings, built either by the Portuguese about the 16th century, or by the Mahrattas when their kingdom controlled that coast and also threatened the Great Moghul as far away as Delhi.

I lay-to about eight miles off, and at ten o'clock that night spotted the flash of a lighthouse between lulls in the rain. It was Dabhol, about eighty miles south of Bombay. My first concern then was to regain deep water, and I beat all that night to the nor'-west-by-west across a westerly wind.

By dawn the next morning the wind had lost much of its force, and the lead showed only seven fathoms. The huge rollers of the open ocean in depths of 2000 fathoms now became rearing giants, screening the wind from *Sheila*'s sails, lifting her dizzily and flinging her towards a shore of rocks and shoals. (A wave must be something like an iceberg, with a greater volume beneath than that which shows above. As the sea bed shelves upwards from deep water the space beneath becomes cramped, lifting the wave higher. This effect can be seen on any beach.) By noon I was in five fathoms in spite of all I could do to get offshore. *Sheila* was under storm canvas, because I had doubts of the repaired rigging holding in the heavy squalls, but she was very under-powered between them.

By mid-afternoon she was among an area of shoals, still about five miles offshore, over which the rollers reared high and top-heavy. They raced forward with tremendous speed, their crests curling over and falling with a thunderous roar, churning up the mud and sand from below.

Some of these shoals I could not pass to windward, and to avoid them had to give way precious sea-room to pass inside them. But one caught us, and as a mounting crest approached I hastily lashed myself to a cleat. *Sheila* was lifted high, tossed and flung like a chip of wood, and then battered and covered by the seething torrent of foamy mud. The fall was like descending in a fast lift, and I waited endless seconds for the sickening thud as *Sheila*'s keel struck. Instead an irresistible suction drew us bodily into the muddy cavern of the next curling sea. I knew *Sheila* could not survive, and remember nothing now except the sharp bite of the rope round my waist as solid darkness covered me, and through it all the sensation of being lifted and hurled away by a gigantic force.

The water drained off cabin-top and decks, and *Sheila* rode upright in the sheltered water inside the shoal. We had been carried right over it.

This fright made it startlingly obvious that unless *Sheila* had more power between squalls we would never clear that area. I took in the storm canvas, working fast, and put on full sail less the mizzen. Slowly we made out between the shoals and I even held on to full sail in the squalls, nursing *Sheila* and praying that the repaired rigging would hold. It was no use reefing down for these squalls—reefing takes time before the squall arrives, and by the time you have unreefed after it has gone, it would have been time to begin reefing again to meet the next. It was all or nothing.

The wind steadied further offshore, and by dark I was reasonably safe though still not back in blue water. Landsmen often think that sailors are quaintly sentimental for loving their ships, but this is not so. We love a person not for their flesh and blood but for their integrity, their reliability under all circumstances. These are the qualities *Sheila* has, enduring qualities, and this is why whatever happens I can never be parted from her.

At midnight I turned parallel with the coast, about eight miles off, and checked my course, in case of currents, with hourly soundings of the lead to keep me in about eight fathoms. I had to keep close enough to the coast to see Khanderi Island light (which marks the entrance to Bombay harbour) through the rain, but not so close as to get into the danger of shallow water again.

The light showed up in the early hours. I hove to, hung a light in the rigging and slept till dawn.

Later that morning I cleared The Prongs, a reef guarding the entrance opposite Khanderi Island, not knowing that I had already been sighted, and that instructions were being issued to shipping and aircraft to cancel the alert ordered for me a week previously. Inside the harbour, a representative of the Royal Bombay Yacht Club came out to meet me in a large launch which, if carelessly handled, could have smashed *Sheila*'s side in with the heavy surge running.

The Indian pilot drew near, keeping far enough off to avoid the force of the surge, but close enough to take immediate advantage of a suitable lull when it came. I held a steady course, and he brought the launch close alongside, Colonel Seymour-Williams stepped on board, and the launch drew clear. Never before or since have I seen a small ship better handled under those conditions than by that Indian pilot.

The Colonel had everything organised, and with a minimum of bother *Sheila* was soon in a snug berth, a watchman on board, and the police, health, and customs satisfied. Within an hour I was sitting back in clean, dry clothes after a hot bath, drinking my first beer for five weeks.

Outside the monsoon flooded the garden, and at the end of the lawn waves flung themselves unrepentant against the stone buttress. I looked out through the harbour entrance from my comfortable chair, and let my mind slide back over 7000 miles to that rainy day a year before when I had sailed out of Lymington. I believed it as I believed the experience of a dream.

Three days later came the news of a Liberty ship driven ashore by the monsoon seas, just north of Bombay, on her way from Karachi. In the privacy of my mind I felt a great surge of affection and gratitude for *Sheila*, which somehow passed beyond her into something greater. Radar, echo-sounding, and powerful engines are not yet everything, nor ever can be.

18

*Interlude ashore—an appendix operation—refitting—
social engagements—red tape—happy departure*

Within the first few days in Bombay I decided to stay in India until the opening of the north-east monsoon. In this I would sail to Colombo, and after it straight to Malaya, hoping to reach there before the south-west monsoon overtook me.

There was some fuss in Bombay after I sailed in out of the monsoon. People were very kind and several invited me to stay as long as I liked, to write my book. I declined these offers partly because I had other things to do first, partly because I had very little money, which makes it sometimes embarrassing to live with people who have much. Even the richest bleat how hard up they are, and so do not understand when anyone really is.

Social life in Bombay was not what it had been, and the glittering evenings at the famous Taj Mahal Hotel were sadly missing, because after Independence the State Government had decreed Prohibition. No government will ever stop people from drinking, just as no government will ever stop prostitution, and stringent attempts to do so only lead to greater evils than when these pastimes are pursued with controlled discretion. Surely Prohibition in the United States has proved that within living memory, and the rackets born then are still existent although Prohibition itself is gone.

All *Sheila's* sails and gear, my books and clothing, were stowed ashore and *Sheila* herself drawn up on slips under cover. I planned to be away only about a month but the future is ever uncertain, and at anchor she would have been in danger of the teredo worm, so voracious in tropical waters. A kind person said he would have everything washed free of salt and dried, and thus allowed my early departure to south India.

Here I found that most of the £300 left in Ceylon to meet needs en route had been spent elsewhere without my sanction. I completed forms demanded by the Ceylon Exchange Control in order to transfer the remaining sum to India, and over the next two months some wretched little bureaucrat quibbled over whether I was indeed going to use the funds for living expenses as certified. Even if I had not, the sum was so small it was hardly likely to endanger Ceylon's economy.

Final sanction only came when I wrote that I proposed to transfer the sum by-passing his office, as I could easily have done in the first place. (A friend had accounts in both Ceylon and India. I would simply have paid the amount into his Ceylon account, and he would have given me a cheque for the equivalent on his Indian Bank.) This instance is typical of so many government controls and bureaucracy generally— the waste of time, the massive regulations, the high taxation to pay the numerous salaries, when evasion is so simple. When at last I received the money it was just enough to pay expenses accrued in the meantime, but not my fare back to Bombay.

During these days, subsisting precariously on alternating periods of credit and isolated articles sold, I met a young Anglo-Indian doctor. He was interested in the health aspect of my voyage and I told him of the consistent stomach trouble and its very enervating effect, stating that it was probably the result of amoebic dysentery. He suggested it might be due to a faulty appendix. This organ had never given me a twinge, but he persuaded me to submit to a thorough examination which confirmed his suspicions, and he offered to take it out.

I knew I could not go to sea again after hearing this, otherwise when next 500 miles from land I would give myself appendicitis whether it existed or not. But at that time I had about forty rupees (£3 odd), and simply could not afford it.

'I'll take it out for Rs.100,' he said, 'and you can pay me after you get to New Zealand.' His trust implied a greater belief in my eventual safe arrival than my own.

'But,' he added, 'you will have to have a nurse for a few days following the operation. She must be paid in cash at Rs.10 a day as she lives on her cash earnings.'

In a fashionable hospital at that time I suppose the surgeon's fee, hospitalisation, anaesthetist, and so on would have come to about Rs.600. This was far beyond my reach, and so I gratefully but with some misgiving accepted the more generous but lesser fee.

There is unlimited work in India for men with such skill, backed by a genuine understanding of its need. These two, the doctor and the surgeon, were struggling not to get patients but to build a suitable operating theatre and accommodation in which to treat patients. They had no government grants, nor help from moneyed missions, and being as then unrecognised presumably all their patients were as poor as myself. They had to work under unsuitable conditions in order to construct better ones, and at that time the new operating theatre was not quite completed.

I was appalled when I turned up with my pyjamas at the address given, and entered the 'theatre' then in use. It was in the back room of an old house, the bricks and rubble of a half-demolished partition wall lay scattered over the floor; long streamers of cobweb hung from the dust-covered rafters above; the tools of the trade lay in a tin hand-basin in a corner, and as I entered were being closely inspected by a mangy dog which had wandered in from the bazaar. To my relief the Matron, a pretty Anglo-Indian girl, shoo-ed it away before it came to a definite canine decision.

On the third day I left hospital to avoid the daily payment of Rs.10 to the nurse. In spite of having left all my private papers and possessions loose in my room at the boarding house, and having agreed to pay a reservation fee, my landlady had put someone else in my room. She knew I was broke and feared for her due, and this was a way of demanding my departure. I stood about feeling sick for three hours while the other occupant vacated the room, because I had nowhere else to go.

Thereafter I received little attention. Sometimes long after the normal breakfast hours a servant brought me a cold plate of bacon and eggs, flabby in congealed fat. Sometimes the dirty breakfast things were left there all day beside my bed. Not one of those Europeans who had made such a fuss on arrival ('Oh Major Hayter, do come and stay as long as you like')—not one came to see me. An Indian came, offering help in the way of food and lending me books, and two other people in the boarding house made me a nourishing soup. Overnight, from headline news as a single-handed sailor and officer of a famous Indian Army regiment, I became poor white trash. An interesting experience.

My wound gave me trouble and kept me in bed. By the third day it had swollen almost to the size of a small football, and whenever I left my bed to go to the bathroom I had to support the sagging sack with my hand. In the early hours of one morning it burst, filling the hollow of the bed with blood and pus, but I was too weak to do anything much but lie in it.

At that hour and in that condition fears are exaggerated, so I knocked on a locked door connecting the next room where a padre was accommodated. I explained the situation and thought it might be wise to call the doctor as I felt I was going to die. He talked sleepily around the point, and finally suggested that neither of us did anything definite until breakfast time. In the morning the servant came in, threw up his hands in horror, and rushed out to report. He returned shortly after with a message from the landlady that I was to leave immediately!

Later in the morning the doctor came with the little nurse who had looked after me. Thereafter for a week, after her daily work or when she was free, she bicycled from the other end of the town to bring me food, wash me, change the sheets, and do for nothing all those things which I had left hospital early to avoid paying her for doing.

Soon I recovered enough to sit up each day and write. These articles fortunately sold and so once more for a time I recovered financial equilibrium. Any social life was out of the question.

People always ask, 'Isn't it terribly lonely?' thinking of the weeks at sea miles from land or another being. Then there is no loneliness because you live in harmony with all around you; but when oppressed by a burden of worry and unhappiness a man is robbed of all power of decision, of positive action, of self-confidence, and in those days I submitted to pleasantly planned insults and humiliations normally unthinkable. That is loneliness.

As soon as I began to recover, there was nothing to keep me in south India, and so to pay my landlady and my fare to Bombay I tried to sell my Leica camera. The Indian merchant rightly assumed that such an article was offered for sale only because of a pressing need, and offered me a fifth of its true value. I refused the offer partly on principle, partly because without the camera any future articles would be harder to sell: it was my livelihood.

That evening I received a letter from John, the Adjutant who had made my life such hell in subaltern days. I had written to him simply recounting that all was not well and that my voyage might end in India. He had replied by return stating that he could cable me £150 whenever needed, and not to worry about repayment until the future made that possible and convenient. The money arrived a few days later, more than clearing me of all commitments in south India, and leaving a surplus after providing my fare back to Bombay.

On arrival in Bombay I called on the person who had kindly undertaken to dry *Sheila*'s sails and look after her.

'Oh hello,' he said, 'I'm afraid I've been so busy I've not even seen *Sheila* since you left.' The sails were dark with mildew, my precious books were mostly lumps of solid pulp, and *Sheila*'s planks had dried and opened so much that she sank like a stone when launched. So I set about the long, monotonous job of re-caulking every seam in her decks and hull.

I have not expected help as my due—if you undertake a voyage like this you must be prepared to stand alone—but when help was offered and accepted I made no other plans for that particular problem. So often by undertaking voluntarily and kindly to do a certain thing and then not doing it, people imposed a greater loss on me than if they had left me to do it myself.

Sheila was due for a major refit in Bombay, and perhaps no one who has not owned a boat will know how much work this entails. In those days so soon after my operation I was not overflowing with dynamic impulse, and without the very great help of the Indian Navy, the job would have been beyond me.

It is not impossible but it is very depressing to live on board among the mess of refitting, and so I found lodgings in a hotel at Rs.8 per day, about the cheapest available above a doss-house. It was situated deep in the dock area, on top of a huge dingy building with storerooms and strange offices in the lower storeys. I insisted on a room to myself and this was no worse than some I had known, but the food was poor—usually a vegetable boiled to a white pulp and the goat meat swimming in ghee (clarified butter). The bathroom was a small bare room with a cold water pipe jutting out of the wall, and the gutters were blocked with all the things the water should have taken away. The latrines were indescribable, being merely slots in the floor over which you squatted.

Soon after my return to Bombay I dined with some of those who had previously welcomed me. After a pleasant evening very much as one might have dining in London, they drove me back to my 'flat', I directing the driver. Conversation lapsed as we entered the dock area, the stench of rotting garbage filled the car, and the huddled, rag-wrapped forms of Bombay's sleeping poor slid past on the narrow pavements until we stopped at the dark hole leading to my stairway.

'For obvious reasons,' I said, my evening shoes ankle-deep in rotting garbage, 'I won't ask you in for a nightcap; but thank you very much for a very pleasant evening.'

'Oh but, Major Hayter, we've enjoyed it so much. We'll contact you soon.' I never saw any of them again.

One great disadvantage, offsetting all the advantages of contacting the European Yacht Club in foreign ports whether English, Dutch, French, or Portuguese, was that thereafter I was committed to the European set, and by receiving kindness I was under an obligation to conform to their way of life. This always prohibited to varying degrees the chances of meeting the local people. I had gone to India to meet Indians, not Englishmen, and with my release from social commitments considered befitting to my previous station, I was allowed to relapse into a more natural existence.

I became friendly with a young Indian who owned a small yacht in which he and his family spent every holiday exploring the fascination of the harbour. He told me that if I had any difficulty getting gear for *Sheila*'s refit, to let him know and he would get it for me. The great sprawling mass of Bombay bazaar stocks everything from brass tacks to a ship's anchor, but the difficulty for a stranger is to find what he wants.

For three days I searched the bazaar for some wire rope for the mizzen standing rigging, and apparently there was none to be had, so I asked the Indian for help. Three days later his 'boy' arrived with exactly what I wanted—it was new but had been used so I asked the boy where it had come from, and he told me from the Indian's own boat.

He had not been able to get the wire in Bombay and so had stripped his own boat, thus putting it out of action for no one knew how long and so foregoing the one form of relaxation in which he and his family found so much pleasure. That is a lot to do for a stranger passing through one's country. I sent the wire back, knowing the cost it imposed, but have wondered ever since whether it was indeed the right thing to have done.

Another Indian, while travelling on business in south India (he owned everything from tea-gardens to hotels, industrial concerns to blocks of flats), had asked me to dinner, but at that time for various reasons I had declined to meet him. I did however meet him in Bombay, and after dining together we spent the rest of the evening talking. As awareness of the other's interest grew, the talk flowed into those subjects which are not peculiar to one race but are common to all. Such talk breeds friendship, which on outward knowledge I had not thought possible between this man and me, and finally I asked a question which had been in mind since his invitation in south India.

'Why are you troubling with me? You couldn't care less for ocean cruising, so why the interest?'

'I grew up in a very poor family,' he said, 'and you know a little of what that means in India. I swore that some day I would be rich, and would devote my whole life to that one aim. Now I am rich, very rich, and have made great sacrifice to be so; but you I think have made very similar and equal sacrifices for an object that is exactly the opposite to mine. Now which of us is right?'

And so we sat into the early hours comparing miseries—he was miserable because he had too much and I because I had too little, yet the money we blamed had nothing to do with it.

Another evening I became involved in a circle of big industrialists. These men were millionaires running huge interests which extended beyond India, and who were themselves at home in New York, London, Paris, and elsewhere. They were talking about the political future of India.

'Well,' said one, 'I am going to vote for the communists.'

I was very small fry in that company but dared to ask why.

'Because it's common sense,' he said. 'Democracy sounds very wonderful and modern India is crazy about it, but democracy means rule by the masses, which means rule by those with the least education and knowledge required to rule. People like me are taxed heavily not for expansion but to pay for the ill-conceived experiments and stupid errors of mass rule. I will soon be using my brain and organising ability for the same wage as unskilled labour backed by the inevitably powerful union. By getting the communists into power I could finance the leaders which would prove my loyalty, be a good investment, and leave me in peace to continue as I am.'

Perhaps he was pulling my leg, delighting in my shocked silence, but I could not be sure. Anyway I have long since given up arguing with communists, mainly because it is impossible to demonstrate spiritual values to those who only recognise material ones.

Ann and Neil arrived. They were to sail with me as far as Goa, and I was anticipating their company with real pleasure. Neil was a shrewd Scot with a Scot's humour, and had done quite a lot of sailing. Ann was intelligent, amusing and very beautiful. These two had been good to me, collecting me in the black days when I was sick and taking me to their pleasant home.

Expenses were mounting, and I could see that the cost of living and of refitting *Sheila* was not going to leave enough to clear me from Bombay. In desperation I cashed in a life assurance policy into which I had already paid over £400 in premiums, and the reward for which was just £101. It is indeed true that the less you have the more you lose, until even that little you have is taken.

The last days before sailing were a frantic rush because the dates of Neil's leave were prescribed, but the greatest delay was not imposed by work but by officialdom. If I had been sailing to a British port I would have sailed without clearance papers, but the next port was the tiny Portuguese colonial possession of Goa, and in such places small officials are belligerently great; to arrive there without clearance papers from the previous port could lead to the impounding of the ship (especially as Goa is a smuggler's paradise), and the expense of the subsequent delay could wreck my voyage. After half an hour in the maze of desks in a great building I found the Indian official to deal with my case. A long silent pause followed my explanation, during which I laid no rupees on the table (partly because they were scarce, partly because I hate giving bribes).

This blatant corruption of Eastern officials so often irritates the West, but as far as I have seen the smooth running of business the world over depends on bribery in one form or another, even if it is only entertaining a commercial traveller. The Eastern official regards his extras in the same harmless light, because where in the West we pay high taxation to be re-disbursed in high salaries (comparatively), Eastern governments pay their officials very low salaries and expect them to make good the balance direct from the public.

After another half-hour the official declared that their file holding the copies of my clearance papers into Bombay could not be traced, and would I please come back at noon. This meant the loss of a whole morning, because by the time I had got out to *Sheila* and started work, it would have been time to come back to wash and change for official pursuits.

At noon my tormentor said, 'There is nothing about yachts in the regulations. You can only be a fishing vessel or a commercial vessel. Which are you?'

After a lot of fruitless explanation I said I was a fishing vessel.

'What is the registered number of your licence, where do you market your fish, and why does a fishing vessel want to go to Goa?'

'All right,' I said, 'I'm a commercial vessel.'

'Then you must produce a cargo manifesto, and invoices declaring the items of your cargo.'

At 1 p.m. the lunch-time break intervened before open hostility was declared, whereafter I went to see the head of the Harbour Control, an Englishman with high connections in the shipping world.

'This is easily fixed,' he said, and made an appointment for me with a head official of the Customs. That person was charming, heard my difficulties and decried the inefficiency of his inferiors.

'This will only take a few minutes, but first please excuse me a moment.'

He did not return within half an hour so I made enquiries. The *Bara Sahib* (head of department) had been taken suddenly ill and would not be in office again that afternoon. So I went back to Harbour Control.

I forget the result of that visit, except noticing that the Indian clerk had been listening intently to the conversation, and when I left he slipped away from his desk and caught me up in the corridor.

'Sir,' he said, 'I have school mate who is clerk in Customs. I show you to him Sir, and he help.' It was a straw to clutch, and having failed from the top I thought it might be worth a try from the bottom.

The wheels began to move, and after walking miles from desk to desk, department to department, at noon the following day the original official said, 'Now Sir, all the papers are in order and only require the Bara Sahib's signature. Please come back after lunch and you will have your papers.'

'Is there absolutely nothing else?' I asked. If the afternoon was going to be free I would arrange other matters connected with *Sheila*'s departure.

'No Sir, nothing else.'

After lunch I called for the papers.

'Have you the Medical chit Sir?'

'Yes. That is already with the papers.'

'No Sir. We have the Port Health Certificate, but another is required from the General Hospital. It is to certify that no member of your crew has received medical attention during the ship's stay in port.'

'Why the hell didn't you tell me this before lunch? And anyway I have no crew.'

'Oh but Sir, the regulations state—'

So I walked to the hospital, fought for a place through the smelly, shouting chaos surrounding the office and finally got attention.

'But you don't need such a certificate,' said that worthy. 'Tell the clerk he doesn't know his job. He's only trying to get a few rupees out of you,' he added, looking at me hopefully.

This is the technique that bent the Imperial knee and it now bent mine; with humility tossing precariously on a hidden seething rage I meekly begged for his illegible signature.

Later in the afternoon I finally collected my papers complete and in order, and that night under a full moon worked on *Sheila*'s rigging until long after three the next morning.

We were not shipshape but it was possible to sail by noon that day, 13 January, 1952, and I had just decided not to delay for lunch ashore when our last impediment arrived—the pleasant skipper of the Shell company tanker *Tornus* which was in dry dock just astern of *Sheila*'s berth. There was much sense in his kind suggestion that we have a quick lunch with him.

'It will save time in the long run,' he wisely remarked.

The pleasant cabin, the skipper's qualifications as a host, and a bottle of good brandy finally released us in mid-afternoon. We were in a holiday mood, it was a beautiful day, stores lay haphazard on the bunks, we squatted happily on the suitcases filling the cockpit, and the dinghy towed astern. It was a good way to start for a picnic on Elephanta but unwise on a 250-mile voyage. However it was pleasant not to be so hellish serious for once, and Neil opened a bottle of beer as a libation to Neptune. This is the only voyage on which I ever drank at sea.

19

*Two passengers—Malabar Coast—Goa—
monasteries and temples—international concord*

It was the period of the Indian cold weather, and the steady north-east monsoon blew serenely, bringing fine days of warm sunshine and cool clear nights. The navigation was easy, the winds fair, and the perfect conditions made the voyage a health cruise much needed after my operation, the harassed months in cheap boarding houses, and the hard work of fitting out. And it was joy indeed to have company.

The hours at the helm passed pleasantly with Ann to talk to, and Neil adding his comments from the galley where he took over the duties of cook, steward, and barman. He and I split the night watches between us, and so there was none of that lack of sleep which had deadened days during other voyages.

I suppose Ann should really have been in the galley but Neil and I had decided to make this a real holiday for her, and anyway the delight of her conversation was more obvious than whatever she might have created in the galley, where my scarcity of pots and pans and the single-burner stove imposed limitations to a woman's way of cooking. Perhaps Neil was creating a good impression for a day when he might have his own boat. So often during my voyage have I seen the harmony of a blissfully happy marriage threatened only by the husband's love of boats, and the wife's, 'It's no holiday for me—I'm pushed out of the way into a tiny smelly little hole, where a horrid stove flares up without rhyme or reason and terrifies me, and I'm expected to produce endless meals and cups of tea for ravenous brutes who talk of nothing but boats, boats, boats. And then Henry says he can't understand why I don't enjoy it.'

We kept about ten miles offshore, where the chance of running into fishing stakes was less during the hours of darkness, and the beautiful Malabar Coast slipped past breathing its romance and history.

Under the chain of the Western Ghats, where the high plateau of the Indian Peninsula falls sharply into the Indian Ocean, are many headlands still holding the squat forts built by the Portuguese in the 16th century. Some of these, on the rocky out-crop of an isthmus isolated at high water, are huge square blocks of massive masonry which when seen beyond the horizon look like coastal islands. They commanded the harbour in which the ships lay, were almost impregnable to attack from either land or sea, and enclosed not only the soldiers and sailors but the merchants, stores, livestock, priests and women. This was in the days of early European expansion when trade was initially carried on from these fortified posts; it was only later that trade was expanded inland leading to territorial conquest and rule. But now the tide has turned.

At four o'clock on the morning of the 16th, three days out from Bombay, we picked up the light on Cabo Aguada, the northern mark of the Rio Mandovi.

Present-day Goa is situated about seven miles up that river, a real pirate hideout river. A grim fort frowns on a rocky bluff above the entrance, and soon the yellow palm-fringed beaches narrow the channel, which winds inland out of sight between steep jungle-clad hills.

I always find it strangely thrilling to sail *Sheila* in a river, perhaps because her movement is more easily seen as the nearby banks slide past, but even more perhaps because I feel I am taking her out of her normal environment into something new, where instead of my being in her care she is in mine. It is a little like taking a car off a main road, slipping through a fence, and driving over virgin land between bushes, around rocks and holes, complimenting the car's capacity to negotiate hazards not normally encountered.

Navigation was certainly different here. Swirling currents tempted *Sheila* towards the banks, and no chart was ever so up-to-date as to show the ever-changing sand-banks; the channel had to be discerned by the flow and colour of the water. The Pilot Book said, 'The channel is marked by red buoys on the northern and western sides, and by black buoys on its eastern and southern sides.' And then it added, 'The buoys described above are not to be depended on, and are subject to alteration.' That was so, because some were high and dry.

Finally we rounded a corner and there was Goa, a small Continental town of square, whitewashed houses, a pink hotel, and cafés along the pavements. In the swift-running stream and along the wharves were a few coastal steamers and several huge 400-ton sea-going Arab dhows, their high poops reminiscent of the Portuguese galleons before them.

Most of these dhows were beautifully kept, the hulls painted or varnished, the blocks and splices picked out in black or white, and the huge lateen sails neatly furled on the long curving spars. The crews were mostly short very thick-set men of tremendous physique and many of their faces showed the mixture of negro blood, probably derived from luscious African slaves ferried across the Red Sea—as they still are to this day whatever the world may think to the contrary.

The inhabitants of Goa are Indian villagers and Indian merchants, who are Hindus; others of mixed Indian and Portuguese blood, bearing such names as de Souza, de Ferdinandez, de Silva, are Roman Catholics; and finally a few Portuguese who are the top administrators and officials.

I was interested to find out the reaction of these groups to a recent broadcast by Pandit Nehru, saying that he would receive his Goanese brothers into the Indian Republic in spite of imperialistic opposition (to his imperialistic desire to do so).

A Goanese whom I asked advised me not to ask questions because Goa is a police state, and if an informer reported my curiosity I could be put into gaol without explanation, and a trial indefinitely deferred. Whether this was true or not I do not know because I stopped asking questions, but from what I gathered and saw, both Indian and Goanese villagers would welcome incorporation into the Indian Republic, because their lot under the somewhat casual Portuguese administration left much to be desired. The Indian merchants were against it, as this would neutralise their trade in luxury goods, imported into Goa duty free and sold to British and Indian visitors from India. I understand the Portuguese simply said, 'Yes, you can of course take Goa any time you wish; but remember that we have similar possessions in China, Africa and Timor, where there are also many Indians. They might be embarrassed if our friendship with India was disturbed.'

After two days in Goa Neil's leave ended so he and Ann departed, to my sorrow, and that afternoon I sat at a café table writing up the notes for this chapter. A young Goanese sat nearby often glancing in my direction, and it was easy to see that he wished to speak with me. Before offering any encouragement I tried to assess what he might be, and from the hungry look in his deepset eyes I decided he was a student indulging in a little hero-worship.

I glanced up and he came over, asking politely if he might be seated, and soon explained that having heard of my voyage he had wanted so much to meet me. He also told me much of Goa, its history, and present-day trends.

'Have you been to Old Goa, or further inland to the ancient Hindu temples?' he asked.

'No,' I said, 'but I should like to very much. How do you get there?'

'I have a friend with a big house in the country beyond Old Goa. If you wish we could stay there the night, after seeing both Old Goa and the temples that day. We can come back the following day down the river by launch.' And so it was arranged.

Some miles behind Goa is Old Goa, once a large thriving city but now in ruins and overgrown by jungle. It was described at the end of the 16th century as 'a place of wonderful military, ecclesiastical, and commercial magnificence', and even today is still the ecclesiastical metropolis of Roman Catholic India. The story is that it was deserted overnight many years ago, whether because of plague or some strange panic I was unable to discover. It now consists of huge almost deserted monasteries, convents, and churches, one of which contains the tomb and body of St. Francis Xavier embalmed in a glass case. He must have been a good and remarkable man to have gained the adoration of so many for so long. No one could tell me just what he had done, and from what I learnt and saw locally it seemed that his crumbling body was more fervently worshipped than the example he had set.

I entered the church to view the withered remains, more macabre than any naked skeleton. Having seen so many of my Gurkha troops sprawled lifeless and grotesque among the bamboos of Burma, and knowing with surprise no pity, I had learnt that there is no connection between a corpse and the living beings I had loved.

From Old Goa we took a long bus ride inland, and visited a very old Hindu temple. We took off our shoes on the top step and entered a long, stone-paved colonnaded hall. Running down the centre was a spacious rectangle, at the far end of which sat a small black stone god. The air was filled with a penetrating scent and a pregnant silence permeated the room, which seemed to set it strangely apart from the brilliant sunshine seen through the outer row of columns.

A priest in a flowing dhoti, the sacred thread of the Brahmans over his naked left shoulder, entered from a wrought iron door which led to a dark recess behind the little black god. There was the inner sanctum ruled by the presence of the temple god, probably decidedly phallic.

Up to a few years ago I would have been allowed to enter there, as I had in the past entered the inner sanctum of other temples, but now it is denied to Western tourists, journalists and their like; too many viewed these gods goggle-eyed, and returned to their world without understanding to publish the abuse and ridicule of their own misguided interpretations. The Hindus were thus forced to deny their religious tolerance, not to shield their gods from slight but to deter the evil which comes from the desecration of another's religion.

The little black god meant nothing to me, being merely a symbol of which I had no need, but I felt the same power in the silence of that Hindu temple, the same peace I'd found in a Mohammedan mosque, the same comfort I was to find later in quiet thought in a Buddhist shrine, the same reverence imbued by the atmosphere of my cathedral in far-away New Zealand, and the same open wonder I had known one early dawn in Notre Dame in Paris.

I can find no word to describe this exactly, awareness perhaps, but more the sense of a vacancy hoping to be filled. It is like the wide-open mouths of blind young birds, waiting for the food they need, knowing the need but not the food.

My companion felt none of all this. He was full of unformed superstition and groping disbelief, being a renegade Roman with ancient Portugal and India in his blood, and only later did he show me the burden he bore. He now asked me if I would like to see the temple dancing girls.

These girls have been known through centuries as Vestal Virgins, temple dancing girls, the gods' hand-maidens, or bluntly as temple prostitutes. In India, from tiny children they learn the intricacies of classical Indian dancing, and it is in the temples that this dancing retains its purest form. They also learn the art of love, again handed down through centuries of sensual experience and investigation. And they also serve those men who have need of them before entering the temple to worship.

The theory is that no man can worship God and perceive His presence while his mind and body are not in a state of peace, of tranquillity. And so before worshipping a Hindu bathes his body and puts on clean clothes; he fasts to a minor degree, so that his mind is lucid and his body not over-burdened; and he may visit a temple girl. Who are we to deny that sensual desire does not destroy tranquillity? It may be excluded by an act of will, which in most of us destroys tranquillity as effectively as that which it drives out; or it may be neutralised simply by satisfying it, just as the desire for food is removed by food. In that state of release incidentals do not anger, nor bitterness nor malice enter the heart, tolerance comes easily and beauty may be seen in all things. And that is the state in which to worship.

We followed a path behind the temple and came to neat rows of simple thatched cottages, a low wall before each enclosing a small garden of flowers. A wizened old woman showed me into a small bare room with a beaten earth floor, and a doorway covered by curtains at the far end. These parted and the girl stood before me in a flowing white sari, strangely submissive, yet giving out a strong impression that sensual experience, though known, had never touched her.

She would not let me take her photograph for my book, and nor would any of the others whom I later visited for that purpose. My companion told me not to worry as he knew an Indian girl in Goa, young and beautiful, who would be flattered to pose.

'You can put her in your book,' he said, 'and who will know the difference?' I agreed.

Later in Goa I met that girl. She was young and comely, was also dressed in a sari, but I realised immediately that I could not use her photograph as that of a temple girl. She gave no impression of being untouched, whereas the others had held the look of virgins; to have used this photograph would have been an act of sacrilege I was afraid to commit.

We left the temple and went on foot along narrow pathways, through untidy villages of small mud houses and accepted poverty, over streams and waterways in unstable dug-out canoes, and arrived with the setting sun at a huge spread-eagled house. It stood in a large compound surrounded by a high wall, the whole set in a vast tract of green paddy fields.

The master of the house had left a note saying that he would be away for another two days, but that the house was ours. The two servant girls giggled a greeting, put a bottle of fiery arak on the table and set about preparing an evening meal. They were just budding, and in their dark slanting eyes was a cautious naughtiness, with caution yet predominating.

I was shown a darkening room where a pair of sandals and a sarong lay on a chair. One of the girls brought a light and led me across a courtyard to a cold bath. On return, two beds had been put in the room hard up against each other, side by side. I took one and pulled it away towards the window.

My host returned from his bath, saw the two beds now apart, and asked, 'Who pulled the bed away?'

'I did,' I said. 'You don't want to sleep on top of me do you?'

And then he said with wistful eagerness, 'It would be nice to be close.'

Then it struck me, and suddenly one or two minor gestures and remarks of the day assumed a significance unnoticed at the time.

'Now listen,' I said. 'In an Army barrack-room beds are a minimum of seven feet apart. It's healthy, and that's the way we'll keep it.'

After an excellent curry supper we sat for a time with the arak, chattering with the girls in some language incomprehensible to me. A few sharp words were necessary before sleeping, and more furniture moving, because on retiring to our room the beds had again been placed side by side. He thought I was unreasonable.

I returned to Goa by launch that day, dismissed my sad companion, and immediately set about preparing *Sheila* for departure. Over the next two days the Portuguese officials became increasingly and unnecessarily trying, sometimes keeping me an hour or more in question and answer, supposedly connected with my clearance papers. Finally came the question, 'Why did you go inland?'

'To get information of this country.'

'What sort of information? Why do you want it?'

'Because,' I said, 'I'm writing a book.'

There was a lengthy discussion among themselves from which they emerged beaming.

'Do you like Goa?'

'Have people been kind to you?'

'Will you write in your book that it's a nice place?' And I departed with my clearance papers half an hour later. I met this reaction in several of the smaller countries during my voyage, and even in some of the larger.

The last night was a riot at the *Café Fiesta*. It began quietly enough until some Portuguese asked me to join them. Even then all was reasonable until one raised his glass, 'You Vasco da Gama, nowadays, *si?*' Then another remembered that Portugal was England's oldest ally; in vain I tried to explain that I was a New Zealander, but this was drowned in the gathering torrent of patriotic fervour and international concord—until long, long past midnight.

I hate sailing on the morning after a long farewell session the night before. Such parties are amusing and often instructive but are better

Adrian around thirty years of age

Sheila in the yacht basin at Algiers

Anti-fouling *Sheila* as she lay against the Red House, Singapore

South-west Monsoon

I lived on Dorado whenever catchable

The beach where I landed in Australia
and tied the dinghy to a bush

...a corrugated iron affair with a falling tin chimney

Adrian after landing at Geraldton

The morning after arrival in Fremantle from Geraldton

Adrian, left, on arrival at Westport, NZ, very thin, with his brother Jim, right

Sketch of Adrian by Maureen Connell for the first edition of *Sheila in the Wind*

Sheila II at her official welcome at Aurora Sailing Club, Nelson, NZ, 1956

Adrian sailing *Sheila* in Auckland, 1970,
for a television documentary: 'Isn't it Terribly Lonely?'

Adrian Hayter in 1970 during the making of 'Isn't it Terribly Lonely?'

arranged several days ahead, because no matter how well you know your ship, the wind, tide, and out-going channel, it is madness to mix drink and sailing, single-handed sailing anyway. There are always many small but important details concerned with a ship leaving harbour for the open sea, any one of which can too easily be overlooked by a fuzzy mind. However in this case hauling in thirty fathoms of chain, and a fifty-six-pound anchor on thirty-five fathoms of heavy manilla, did much to remedy the devastating effects of international concord attained only a few hours previously. What a release it was to regain the open sea.

Over the lovely days south from Goa I sailed closer inshore than previously, to see more detail of that glorious coast. It was a straight line of long yellow beach stretching for miles and miles southwards to Cochin, fringed with green palms, and the bold rampart of the Ghats standing hugely through the haze beyond. And there was nothing to worry about except the gentleness of the wind, but even that was no worry because I could not change it.

I usually moved offshore during the afternoon, and with the southwesterly breeze which often blew in defiance of the season *Sheila* sailed herself to the south through the cool clear nights, allowing me hour snatches of comforting sleep below. If I did continue inshore to catch the night breeze, I stayed awake at the helm because of the numerous dhows on that lane for the same reason, but nearly all going north. They carried no lookout forward nor any lights, but the creaking of their blocks and gear came through the darkness before them, warning me in time to avoid their phantom shapes. I carried no lights either, considering them an unnecessary expense, as also presumably did the Arabs.

Night came to enshroud this great continent where I had served as a soldier, in another life it seemed, and now it was hard to decide which was the dream. I always stayed awake from about four in the morning to feel each day the ever-new wonder of a dawn gently pushing aside the night, to hide the stars and turn the black water into the greens and blues which came with the sun. Each morning the smoke rose from the fishermen's cottages in the fringe of palms, and again India lived peacefully under its thousands of years, apparently as unchanging as the eternal truths it sought.

In those days, too, perhaps I also drifted outside of time, reminiscing of India's search and the thoughts she had set in motion.

I was not concerned with Hindu ceremony and ritual, nor with the many forms of worship practised by a wide variety of sects. Sects is not quite the right word, because they appear more as groups differing not so much in belief as in the paths towards it. Some of those paths have gone astray through the centuries, and the original aim has become blurred by distractions found en route, or by the human frailty of a priesthood unable to bear the divine powers it may have claimed. All these things are bound in time, they come and go, chop and change, and I was seeking outside time to find that which has no change.

And I have written many of those things down (purely for myself) to capture and hold those thoughts which otherwise are often lost; some had not come then and still lay in the future. Over much of my voyage my mind was like a child standing on a sweep of garden lawn, clutching at autumn leaves passing in the wind, stretching a hand for one, leaving it for one yet closer, running after another, stooping to capture a bigger, rushing forward to others more golden. They all came from the same tree.

20

Arrive Cannanore—the lost anchor—departure

About 250 miles down the coast south of Goa lies Cannanore, and I decided to call there merely because I wanted to. This was a part of India always talked of in romantic vein, and was outside the normal range of military cantonments and the efficient bustle of imperialism doing good; it was a part of India we in Gurkha regiments never saw.

Dawn brought to light two fishermen sitting in a frail canoe moored to a set-line. They had evidently been out all night and were about twelve miles offshore, pinched by the chill moist breeze and shivering under the threadbare cotton wraps which barely covered their nakedness. They shouted something but only after I had passed did I recognise the one word 'cigarette'.

I would wish it on no man to be in such a situation without cigarettes, so went about, ran into wind alongside the canoe and passed them a half-full tin of 50. One took a box of matches and his hands felt icy cold, but as I hauled *Sheila*'s jib aweather to swing her bow clear of their set-line, they were already inhaling deeply and blowing the smoke luxuriously through their nostrils. The next half-mile brought their fading cries, *Salaam Sahib, Salaam Sahib*.

At dusk next day I passed round the old fort guarding Moplah Bay in which Cannanore rests, with thoughts of the Vasco da Gama whose name linked with mine had brought such joy and tribulation in Goa; he too had rounded this point under sail nearly five centuries before, in fact only a few years after Columbus had sailed for America. The anchorage was wide and shallow, giving a heavy roll to the swell from the outside ocean, but the bottom was good holding ground for the CQR anchor. I let it go in three fathoms on fifteen fathoms of chain.

It is always a pleasant feeling, after arrival at an anchorage after dark and making all shipshape above deck, to go below to a lighted cabin. It is home, in spite of your wanderings. While the supper cooked I took bearings on two points ashore, and entered these in the log. Before going to sleep after supper, I checked these again to ensure that *Sheila*'s anchor was firmly embedded in the soft carpet of sand beneath the dark waters.

As I was washing up the breakfast things early next morning, a canoe drew alongside with a wonderful old man at the steering oar. His face was wrinkled like the inter-lacing of a net, telling the years' story of sadness and hardship defeated by faith and humour, to leave a wisdom and tolerance making his acquaintance a gift to anyone. Of the other two men one was nondescript and leaves no memory; the other was short and muscular and looked like a sea-going Arab.

The old man spoke Hindustani, and he explained that they had never seen the type of *Sheila*'s hull and rig on that coast. He drew out my story, translating it to the others, and all were so genuinely interested that I asked them on board. They were enthralled with the galley and cabin, but explain as I would they persisted in thinking the ship's lavatory was some sort of engine. Why talk of soil disposal on a small ship set in a vast ocean?

Before leaving the old man thanked me and asked if I would like some oranges, which were plentiful ashore and the best in all the East (according to him), and because it is sometimes better to accept a gift than to refuse it, whether you want it or not, I thanked him and said I would. It is pleasant to meet the unspoilt native, far from the beaten track of commerce and corruption.

The canoe returned half an hour later with the two men and six beautiful oranges. I accepted them gratefully, and the nondescript person asked for two rupees eight annas! This price I knew to be absurd, even in Bombay; it is strange this trait in human nature, whereby we may not mind giving two rupees eight annas (a rupee is about 1/6d.), but as soon as we feel we are being exploited we fight for every penny. I eventually settled for one rupee, and learnt later that the bazaar rate was half that. The tragedy was that this incident destroyed all the previous pleasure of their visit.

It was in Aden that I had met and been friendly with an Arab through similar circumstances. He spoke English and told me he had worked on merchant ships on voyages to England, South Africa, America and Australia. Later he had done some task for me, and to preclude the chance of recriminations I had first obtained his agreement to a payment of ten rupees. On conclusion of the task he had shamelessly demanded fifteen rupees—and I blew up.

'You ruddy people are all the same,' I raved. 'You make an agreement, you do the job, and then demand half as much again.'

'Yes Sahib,' he said quietly. 'Same too America, England, Australia. I been.'

In the afternoon I went ashore to buy a few supplies, happy to be on land but glad of the thought of sailing again next day. I felt terrific—the preceding days of easy sailing and undisturbed thought had brought a feeling of well-being, a sense of powerful awareness and an exultant wonder at all nearness; it also inevitably brought an equally powerful awareness of all that our physical life may give, readily obtainable in Cannanore. And so against the nearness of happiness was the nearness of temptation to destroy it, as I always had done in the past.

This becomes a raging devil which is one moment banished by a deep wonder, and the next rides in on the sheer brutality of the physical need, sweeping you away in a torrent of desire and splitting the mind in two. One reminds you of the past with its glimpses of extreme happiness, its repeated failures and the misery they brought; the other says—well it says much, all nonsense, all convincing.

I mounted the steps of a shop to ask if I could leave my parcels there to be collected in a couple of hours' time, and a strong feeling of premonition possessed me. I cast it aside but as I laid the parcels on the desk explaining to the shopkeeper the feeling returned urgently.

All it needs is a simple act of will, a simple step through the barrier from one mind to the other, but it must come with the thought that prompts it because with hesitation it is lost. I grabbed my parcels, leapt into a passing tonga (pony cart) and urged the driver towards the harbour.

Sheila did not come into view as the sandy road curved to the beach, but on the fringe of sand I flung the driver two rupees, hauled the pram into the surf, and rowed with steady sweeps. A mile away and just outside the long wide curve of white surf *Sheila* rode the mounting rollers gracefully, her bow pointing seawards, and I knew how little water there was beneath her.

My arms were almost numb with aching as I clambered on board, made fast the pram, feeling sick with the exertion, my own disloyalty and the probable loss of *Sheila* because of it. Her keel thudded on the sand as each swell dropped her, sending a shiver all through her to the very top of the mast.

It took only a moment to get on sail, and as *Sheila* filled and slowly gained way on the starboard tack I went forward to haul in the anchor chain. A link had broken about four fathoms from the anchor, fortunately leaving the remaining eleven fathoms to act as a drag, and this had given me the time in which to save her.

I let go the fisherman anchor in three fathoms. There was no devil in me now—he had been as deflated as myself, and I wondered whether he and I were not one and the same.

The lost anchor had to be recovered if possible, because I could neither obtain nor afford another in Cannanore and it is unwise to sail with only one. I looked up the original bearings of the anchorage in the log, took the hand-bearing compass and rowed out to where the bearings intersected. The depth was only three fathoms, not too deep for me, but the bottom could not be seen through the heaving sandy water. I needed help—it is unwise to dive alone in tropical waters so far offshore, and the light pram would have drifted away while I was under water. So I rowed the half-mile to the shore where the village straggled down to the beach; in the first house I approached was the old man with the wrinkled face. No one else spoke Hindustani.

The old man said there was a famous diver in the village, but my hopes sank when they brought him to me, He was a small, very skinny being with rheumy eyes, in his tropical middle-age, which is the equivalent of about our sixties. However he was keen to try, only unwilling to believe that I could drop him near enough to the anchor to give much hope of finding it in the swirling sand caused by the swell.

We all clambered into the pram, and I rowed out on to the first bearing and then along it until the second intersected. Over went the diver, his bony chest covered in wrinkled worn-out skin faintly moved to maximum expansion, to capture about a cupful of air, and down he went. My hopes went with him.

Almost immediately there was a splashing and spluttering ten yards astern, a skinny arm flayed the water as if in resistance to some sea monster trying to pull him under. I hastened the pram to him and the old man grabbed an arm.

'All natives exaggerate,' I was thinking. 'He can hardly swim, let alone get to the bottom.' And then I saw the end of the chain in the old man's hand!

The compass bearings had been dead accurate, but how that miserable mite of humanity had managed to lift three fathoms of chain to the surface, and hold it, I would never have believed had I not seen it with my own eyes. We were all shouting and laughing with the success of the combined operation, until the old man was in haste to return to the village and tell the tale. There it was told over and over again, the anchor inspected, my compass inspected, the skinny diver dancing on the sand in pantomime, clearly showing ever-increasing depths, thickening clouds of swirling sand, and then his tiny cupful of bursting air dragging hundredweights of massive chain to the surface. The old man gave me a mug of some dreadful drink, which I sipped happily while remembering an extract from the Pilot Book. 'Small quantities of water, which must be boiled and filtered, may be obtained from wells on the beach.' To hell with dysentery!

That evening I dined at the small local club with a Colonel and Mrs. Marsden, two kindly people whose company was soothing after the turmoil of the day. And early next morning two loops of rope made a cradle for the pram, which I then lifted on board with the main halyards and lashed down on its bracket over the skylight; hauled *Sheila* up short on her anchor, got on sail, and as she gathered way I took in the slack of the rope, snubbed it, felt the tug as *Sheila* broke the anchor from its sandy bed, and as she sailed herself out to sea I got it on board, lashed it down, coiled the halyards, returned to the helm, and set her on course for Cochin. Cannanore will be remembered.

21

Arrive Cochin—depart Cochin—the biggest whale—arrive Colombo

The glorious weather held and during the four days' voyage to Cochin, about 150 miles south of Cannanore, I held well offshore to avoid the coastal dhows at night, and because the open sea drew me there.

A normal procedure at sea before arrival in port was to compile a list of all that must be done there, and this included everything from details of maintenance to making out a stores list for the next voyage (the charts and Pilot Book covering this had been bought in Bombay). I wrote letters of thanks to those whose names and addresses I could remember, people who had helped or been kind to me in the last port. The letters were stacked neatly in the 'office' and posted on my next arrival. And I carefully studied the current Pilot Book for details of entry into Cochin harbour.

This harbour is almost an inland coastal lake, wide and beautiful with the entry of numerous rivers and waterways, all fringed with palms except where the main town stands, yet showing distinct traces of former Dutch ownership. A dredged and buoyed channel, about 100 yards wide and four miles long, leads from seaward through the shallow water and coastal shoals into the harbour proper. The Pilot Book said, 'It is advisable for Masters of all sailing vessels to employ the services of a Pilot… should they not do so, extreme caution must be exercised when entering the harbour.'

This remark when read in Bombay had induced me to buy a large scale chart of Cochin, on which the entry did not look so difficult.

I have always disliked the idea of handing *Sheila* over to a pilot because, although these men are professional seamen, none as a stranger could know *Sheila* as well as I do, and although control is handed over to the pilot, responsibility as master stayed with me. It seems unreasonable to hold a responsibility and then hand over the control needed to fulfil it; and for a ship of *Sheila*'s size the Pilot Book and a large scale chart give all the information that is necessary. But when sailing single-handed these details—details of tides and under-water dangers, buoys and what they mean, channel markings, bearings by which these are picked up and the channel followed, signals to be expected and the location of the signal station, local rules for shipping, where to wait for pratique—all these details have to be studied with care and memorised. Sometimes a ship entering a harbour, particularly a sailing ship, demands a lot of handling and continual attention, not only of navigation and sail handling, but mainly in transferring the memorised details of the chart into visual recognition 'on the ground'. This leaves no time to re-open the Pilot Book to look up some neglected but essential detail of entry.

Cochin Light showed up just before midnight on 3rd February, about fifteen miles off, but the wind was no more than a faint breeze. I slept in hour snatches as *Sheila* idled towards the channel entry, where we arrived at dawn. I anchored on the sloping bank of a shoal just south of the outer buoy to have breakfast, and to allow an Indian Navy destroyer to depart.

The destroyer (No. 110) cleared the outer buoy and unexpectedly called me up by lamp, asking, 'What ship? Where from?'

I hastily grabbed the Red Ensign flying on its staff at the stern and morsed back, '*Sheila II*, England.'

She was the INS *Rajput* on her way to meet HMS *Vanguard* which was carrying Her Royal Highness Princess Elizabeth to Ceylon. This was the second time I had just missed a Royal visit on my world cruise, and is one disadvantage of not being able to keep to a schedule.

A three-knot tide was running out of the long narrow channel (which lies almost due east and west), and the westerly breeze only just held *Sheila* against it under full sail. The engine was too unreliable, in which state it is more worry and danger than none at all.

The conditions were ideal for the spinnaker, and if the wind became flukey in the narrow channel where it entered the land I would simply have to drop sail smartly and anchor against the tide. Anyway I feel in a situation like this when there is wind that it is a desecration to drive a sailing ship with an engine. Think of the centuries of affection, care and skill that men have lavished to create the curve of sails and the beauty of a sailing hull. These come into their own when used with the care and skill that has created them, and are what gives a ship her soul. Such a ship is known as a lucky ship, and she will carry you through dangers that no other quality can. Yet entering a strange harbour under full sail and spinnaker hardly seemed to conform with the instruction to exercise extreme caution.

Again followed moments of this voyage which will never be forgotten. From my height not much above water level the long straight channel reached to the horizon. I could feel the thrust of the sails against the strong tide, and the huge ballooning curves of the spinnaker always bring a strange thrill. We swept serenely up the channel at over four knots, but each pair of marker buoys passed us at not much over one, the passing tide giving the unreal impression that they too marched with us.

Both the outgoing tide and the ingoing wind increased in the narrow entry, where palm-littered spits of land almost enclosed the harbour. With a loud gurgle at the bow we passed inside to the wide expanse of calm water, and as the wind still held I continued straight on towards the palatial Malabar Hotel, until I could plainly see the tables on the lawn, and people raising their glasses to drink as they lounged with their idle conversation. It looked delightful but that had to wait. I hauled hard on the port jib sheet and eased *Sheila* slightly off the wind, letting the main and mizzen take her slowly round to port as I went forward to the mast. As the wind began to spill out of the spinnaker I let go the halyard, and the backed jib held *Sheila* hove to while I gathered in the sail and got the long spinnaker boom inboard. A harbour launch came alongside and gave me clearance-in, and an Indian Naval boat arrived to escort me to a snug berth in the Naval base—INS *Venduruthy*.

The destroyer which had signalled me outside had advised them of my pending arrival.

Over the three weeks I spent in Cochin several items of *Sheila*'s Bombay refit were completed up to standard, and again she was slipped and two coats of anti-fouling applied below the waterline. It was less than a month since she had been anti-fouled 700 miles back in Bombay, but I knew that slipping facilities for a keeler in Colombo were not good, and thereafter lay the 2000-mile voyage to Malaya. I cannot stress too much the need for every care against the teredo worm in tropical waters, and port facilities ahead always had to be ascertained and slipping anticipated if necessary.

The Indian Navy gave me the greatest help, from providing a cabin ashore and making me an honorary member of their wardroom, to putting in many hours of work during holidays and their non-duty hours. On the whole I think that India treated me the most kindly of any country visited—talking of Indians, that is, and British service personnel still in India at that time. So on 26 February, 1952, I said good-bye with a mixture of regret and gratitude, and sped down the long straight channel with the ebb and the engine going. There was no breath of wind.

It took a week to cover the 400 miles to Colombo, during which a noticeable change in the seasons was apparent. The perfect weather of the north-east monsoon was drawing to a close, and the transition period separating it from the next monsoon (the south-west) was setting in. The winds were more variable both in strength and direction, calms more frequent, and huge cumulus clouds climbed in solid white towers thousands of feet into the blue.

Water spouts were common on some days, tearing me between a desire to see one really close and the fear that I might do so. They formed from baggy sacks which hung from the base of the clouds, sending out first a slim pointed finger of dark vapour which slowly grew and thickened, curving downwards towards the sea, which at the last moment rose upward in an urgent mound to meet it. Thereafter the column thickened further, and like a spiral of heavy rope united cloud and sea. Sometimes there were several of these waltzing around the horizon, apparently harmless enough from a distance but probably fatal to *Sheila* had they struck.

One day in the Gulf of Mannar, between Cape Comorin and Ceylon, I saw the biggest whale encountered during the whole voyage. The wind was on the quarter and a pleasant sea running under bright sunshine when the vapoury plume of a whale spouting showed about half a mile away on the port bow. Then he travelled along the surface at about eight knots on a course to cross my own. His vastness was apparent and I held my course to get a closer look, until he back-pedalled when only thirty yards off *Sheila*'s bow, lying on the surface with the waves breaking in a series along his back as they do along a sand-bar.

Unfortunately cinema posters and newspaper headlines have drained the meaning from our superlatives, so I can only say that this monster was huge. In those moments of some apprehension mixed with wild excitement I tried to achieve a dispassionate measurement, and concluded that this whale was about three times the length of *Sheila*, about ninety feet long.

As I drew level he sank below the surface, his shadowy bulk still visible, and swam directly under *Sheila* not much more than a fathom beneath her keel. By leaning over the side I saw him plainly, and the great width of his shoulders seemed to equal *Sheila*'s waterline length, twenty-four feet, but perhaps this was not so. As he cleared *Sheila* on his way the rough water became smooth with the curling eddies and swirling currents thrown up by his huge flukes. I could not help thinking how inconvenient it would be at that time and place, if he had decided to be as unpleasant to me as some of my species were to his.

The entry to Colombo harbour presented no difficulties, as I remembered it well from several previous visits in liners and troopships. A pilot came out, even so, as I passed through the breakwater, merely for the ride I think, and in the gathering dusk helped me to a safe anchorage just off the Royal Colombo Yacht Club.

A dinghy bumped alongside almost immediately, and a voice said, 'Here, grab hold of this,' and a hand rose out of the darkness, beautifully disfigured by a foaming tankard of ale. I grabbed!

22

The Yacht Club—the yacht California—another hard decision—departure

The Royal Colombo Yacht Club appeared to me to be almost ideal as far as yacht clubs go. Although lacking the wider natural facilities enjoyed by others elsewhere, what it did have was organised to the best advantage and any lack in no way depressed the enthusiasm of its members. The Clubhouse stood on a green lawn on the edge of the harbour, the whole front of the upper storey being a wide verandah overlooking the sea; and beneath were the changing rooms, showers, storerooms, members' lockers, and a shed for sails and boats' gear.

None of that is unusual, and what impressed me first was the enthusiasm; members sailed in all weathers, on schedule, and never once did I hear confusion regarding starting times, courses to be sailed, nor of any other essential detail. The second thing which impressed me was the balanced harmony between this enthusiasm and the sensible social aspect, and this harmony is not usual. I put it down to the effect of lady members.

In a purely male club directed to the one purpose of sailing, particularly racing, the undiluted male enthusiasm always breeds a degree of fanaticism, and inside around the bar the conversation inevitably ranges round the one subject of sailing, with the exception that 'drink' or 'women' sometimes pass like flotsam on the verbal tide, but soon sweep beyond and are forgotten.

I appreciate all these subjects, but when pursued relentlessly without distraction they become incredibly monotonous. Like any fanaticism unleavened, the pursued soon rides the pursuer, and sailing in the Club soon spreads to sailing in the home, to business offices, to casual meetings in the street, until the mind is possessed to the exclusion of all else. The fanatical soldier seeking efficiency, the golfer, the bridge-player, the economist, they are all the same and destroy the attraction of the god they so arduously worship. I once became so keen on golf I gave it up.

A club including women members runs the dire peril of becoming a loosely-knit social gathering, where the main purpose becomes obscured, and the detailed organisation necessary to fulfil the enthusiasm for sailing breaks down; as soon as this happens of course the keen sailing members leave or become disgruntled, and the club becomes a yacht club in name only.

This need not be so, and if affairs are carefully directed lady members seem to breathe a sense of proportion into the sailing activities outside, and to exert a broadening influence on conversation inside. At all events, a mixed club is far more pleasant for a visiting yachtsman.

The morning after I arrived I was working aboard *Sheila* when three young Ceylonese rowed around in a dinghy, taking photographs. I was busy and carefully continued so, because a friendly good morning to many such seekers brings the request to come on board, and my subjection to long talks as to how they would do the voyage.

After several moments I was able to perceive that they were not the brash kind, and had come shyly in a genuine desire to satisfy interest and offer hospitality to a stranger—hesitantly, because they had learnt that some whites tend to regard this as a presumption! So I asked them on board, and this led to a friendship I shall not forget. In due course they took me to their homes, introduced me to their parents, and over long conversations taught me much of Ceylon and even more of the Buddhist religion. They talked of this naturally as an everyday subject in their lives, and made no claims that in it lay the only salvation of the world.

One family of another section of Ceylon people took me to their home. They were Burghers, who were proud of their European blood (much of which contains the bluest of ancient Portugal), but they mingle harmoniously and fulfil an important role within the community. I sensed none of the antipathy which surrounds the Anglo-Indian in

India, although much of that had departed with the contrast unconsciously provided by British rule.

The father wore European clothes, had humour and a quiet dignity; the wife wore a sari, and behind her easy-going maternal kindness she was very much queen of her domestic domain; and the daughter was a tall girl of about eighteen, with dark eyes and finely moulded features. She too wore a sari, whose simple grace as worn by an eastern woman is more beautiful than any Paris creation.

These good people were among those who eventually saw me off, the mother giving me a well-packed parcel of luxuries for my voyage. At that time too my other Ceylonese friends gave me a volume of Shakespeare's complete works, and a book on Buddhism by Edward Conze.

About a week before I was due to sail, a beautiful three-masted schooner-barque glided into the harbour and dropped anchor just astern of *Sheila*. She was the *California*, three years out from Los Angeles with four young Americans on board, plus 'Scupper', a ferocious-looking but amiable Boxer dog, which later committed the grievous social error of killing the club cat. The Americans had been cruising in the Pacific, and were to continue via the Red Sea and Mediterranean and across the Atlantic to their home port.

I later met the crew and spent much time with them, and we swapped information of our respective past routes which then lay ahead of the other. They were all first-class seamen and ran their ship on the unique lines of changing the skipper by daily roster. For that day one man was skipper and carried the authority which that post must hold for the safety and well-being of the ship and crew; on the next and following days he gave the obedience to others that he had received from them. Such a system could only have worked where each had the full confidence of all to fulfil the responsibilities demanded, and it was in effect the perfect democracy, where all shared in the leadership and all bore the responsibility for it, attaining the efficiency of totalitarian control but avoiding the loss of individuality which total rule demands.

As the time of my departure drew near I faced again only too fully the problem faced earlier in Aden. I could not wait in Colombo for the next north-east monsoon, which in any case would be a consistent head wind of about force four, an unsuitable condition for a 2000-mile voyage. The coming south-west monsoon would be favourable in direction, but its fury already experienced had left little desire for repetition. The only alternative therefore was to sail then in the transition period, in spite of the doldrum belt lying directly from Ceylon to the northern tip of Sumatra along latitude 6 degrees north. It was also the hurricane season, but the hurricanes would rage in the Bay of Bengal and seldom came below latitude 10 degrees north, and so they forced me to plan my course near the equator.

All who have read of the old sailing ships will know how the doldrums were dreaded: the flat heat-laden calms often lasting for days, the variable puffs of wind demanding all sail to get the best from every vagrant breath, and the sudden tropical squalls of hurricane force under torrents of icy rain, demanding speedy reduction of sail to save the ship and her rigging.

Those ships always aimed to cross the doldrum belt at right angles, a distance of about 100 miles, but now I planned to sail nearly 2000 miles along its length. North of it lay the hurricanes, a risk I dared not take, and south lay the equator, with conditions no better. Everyone said that this route was impossible, but no one had ever tried it to prove this to be so; and even if they had, one failure doesn't prove another. Anyway, I knew if it wasn't possible, then the south-west monsoon would eventually overtake me in mid-June and hurtle me into Malaya.

All this is one of the penalties for sailing round the world the wrong way, but I had already fulfilled one minor object of my voyage, to re-visit India, and I still had another, to re-visit Malaya and my regiment.

And so on 27 March, 1952, I sailed from Colombo for Penang, with five weeks' supplies and twenty gallons of petrol, limited as usual because it was all I could afford.

23

*The rat—the doldrums—a dream—the Monsoon breaks at sea—
Siamese interlude—arrive Penang*

I passed round the southern end of Ceylon three days later, about eight miles offshore, with the first good wind since leaving Colombo. The southern horizon was a city of tossing sails and spars, the myriad fishing craft from the coastal villages.

At midday an awesome bank of clouds marched up from the south, the wind freshened, and the fishermen lost no time in beginning their flight for home. Over the next two hours hundreds of craft, from long slim canoes to small streamlined dhows, all under billowing lateen sails, surged across my course, very skilfully handled as they ran dangerously fast before the following seas. This meant I had to keep out of their way, because if they had turned sharply either side off wind at that speed they would have lost control and broached-to.

The massed clouds looked as if they were just waiting to release all hell, and the men of the last dhows, seeing me hold my course to the east as they sped to the safety of their harbours, shouted and waved frantically for me to do the same. But *Sheila* was safer in the open sea than in seeking safety, and we had to take whatever came.

That evening my rat made his first fleeting appearance. It must have come on board in Colombo when *Sheila* lay alongside taking in water. Little did I know then the impositions this elusive beast was to impose, and only later did I thank heaven it was not a pregnant female. Ten days afterwards I found it had entered the locker in which my cigarettes were stored, packets of twenty wrapped in moisture-proof cellophane, where it had torn four packets into a filthy mess of shredded tobacco and nibbled five others to be destroyed by moisture—a total loss of nearly two hundred cigarettes which imposed great anguish upon me later.

The brute demanded extra work out of all proportion to its size. All food normally left open, such as butter, half-used tins of milk, biscuits, etc., had to be sealed away after each meal. There was no container large enough to take the potatoes, and months afterwards when re-fitting in Malaya I found half-eaten pieces stowed away in inaccessible places of the ship. I could appreciate the brute's desire to survive, and could accept his taking one potato a week, living on it, and then another, but this extravagance was a direct threat to my own survival. And yet it was shrewd foresight on his part, because later when I systematically imposed a starvation diet by shutting away all possible rat-edibles he then presumably survived on his hidden reserves. The rain gave him water. At this time, too, he retaliated. Every morning almost without fail I found his insults in the cutlery drawer, which then had to be sterilised with boiling water. Mistrust coupled with our mutual need to survive forbade our mutual need to co-operate, and placed any tendency towards friendship out of the question.

For the next six weeks, when the south-west monsoon overtook me, I lived in the doldrums. The conditions allowed no routine by which to conserve endurance, no routine of meals and sleep which had helped so much on previous voyages, and the continual sail handling between full canvas and bare poles imposed irregular hours through the twenty-four of each day and night. Yet in spite of all the work I averaged only twenty miles a day.

Flat calms prevailed, when the whole ocean lay to all horizons like a huge pool of mercury, motionless. The heat was appalling, and often around noon even if there was a useful wind I had to seek shade below to avoid fainting in the open cockpit. Puffs of wind did come, stealing in from the horizon, darkening the glassy surface like a liquid stain. It was impossible to forecast how long they would last but each breath had to be used to the full—sometimes to carry me only a mile, sometimes twenty, but always it meant climbing out to the bowsprit end to set the big genoa, or handling the eighteen-foot spinnaker boom to send those billowing curves aloft and give *Sheila* the power we needed.

At any hour of the day or night a towering cumulo-nimbus rode beside us, or crossed the bows, or bore down to snatch at *Sheila*'s masts and rigging, hurling her over to her gunwale, and drenching all in a tropical downpour. And then, even though closely reefed, the jerky quivering of her mast sent shudders from one end of the hull to the other, and I struggled desperately on the steeply sloping deck and in blinding rain to take in what sail there was before the rigging gave. No wonder mutiny was most common on the old sailing ships when in the doldrums.

The nights were naturally cooler than the days, although even at night my body was always streaming with sweat. To avoid the stifling cabin I usually slept on the side-deck, a stanchion half-way along my length so that only one half of me could roll off at a time; sometimes I curled in the cockpit, sometimes stretched along the foredeck.

Six weeks under these conditions sounds like hell on earth, but the closely written pages of my diary tell that the days and nights were crowded. There was a tremendous amount of life in the ocean around me, much of which sought *Sheila* either for companionship or for shelter in the shade of her hull. This is taken straight from my diary:

> It has been a calm day and I spent much of it watching fish—the shadow of the hull pierces the sunlit water like a shaft, in which I can see fathoms deep through the warm rich blue. Four small fish, dark blue, live under the turn of the bilge, and only dart out occasionally to devour some defenceless body of plankton. A bigger fish also lies there, a pale brown with a bright yellow tail edged in black—he's been with me for days and pops out to investigate every cigarette butt I flick overboard. Far below, about 5 fathoms down, is a tight shoal of about 100 fish, about a foot long, blue with bright yellow tails, who cruise together under some directing command and are occasionally chivvied by the dorado—these are as usual everywhere, roaming, hunting flying fish, always

inspecting my dangling toes, and often lie motionless to sink slowly like a gliding plane far down into the depths. And today a newcomer—about 9 inches long, dark smoky brown with blue circular rings over his body. He has a tiny white-lipped mouth and tiny side-fins; he gets about by waggling his very developed dorsal and ventral fins which move him forward, over on his side, or even into reverse.

It's been such a beautiful day—the blue sky, the white clouds, and the molten mirror all stretching into eternity. From the point of view of sailing, of getting from Colombo to Penang, it's been appalling; from the point of view of living it's been heaven.

Another entry shows some of the contrast, and boredom is impossible with contrast:

A bloody night—calms, big swell, thunder storms. This weather is a bastard—burning hot one minute and pouring with sweat, next shivering with cold and drenched with rain. Do sunny days and starlit nights with gentle winds exist?

That magic box, the radio, finally died three weeks out from Colombo, denying me thereafter the time signals of GMT necessary for longitude—and travelling east and west this is important. I had a rate worked out for my wrist watch, but later when the sights differed greatly from DR, I could never be sure whether this was due to wrong time, or to a change in the current which should have changed long ago and was unpredictable in the transition period.

Much of my time I spent reading. The book on Buddhism given me by my Ceylonese friends added much to what they had told me. It appeared that Buddhism was similar to Hinduism, and was merely a breaking away from the latter which had become petrified by an established priesthood, behind whose imposition of dogma and ritual the application of the basic truths to everyday living had become bidden or distorted. So the Hindus also had their Pharisees.

I had once told a padre how a reading of Hinduism had helped explain much of my own scriptures, and asked him what he thought of Hinduism as he had already claimed to have studied it.

'Hinduism,' he said, 'is the most stinking iniquity ever conceived by the human mind.'

How could I find humility and peace and the way to love my neighbour in my own Church, when it spoke with hatred and ignorance, harshness and intolerance against its neighbours? Neither did our Pharisees depart with Christ.

These thoughts came and went in my mind in the vast emptiness of ocean, tearing me apart.

'All this solitude has been too unnatural; it's driving you crazy.' Or else, 'This solitude lets you see things denied without it.' I wrote at the time:

> It is almost as if I am deliberately prolonging my isolation in this vast waste of confused seas, burning suns, icy downpours, and towering clouds—I want these alone to see if all I hope for will come; there's no hope for me on land, where to think like this is madness.

Yet it was the solitude that brought peace, and one night I wrote:

> It's happy sitting here in the cabin, the lamp on the table, the dark shining wood, my books, the chart with its criss-cross lines on the table, the two sails genoa and spinnaker hastily furled on the settee opposite—yet outside this light is miles and miles containing not one other human soul.

And feeling that wider emptiness outside I left the cabin to sit in the cockpit.

The night was beautiful as perhaps only a tropical night far out at sea can be, bringing a sense of deep relief after a scorching day. The line of the horizon was hidden by an invisible veil, and as I gazed over the side the surface of the water was so still and dark as not to exist, and in the depths were a million stars as still and as far away as those above me. I felt if I rocked the boat or looked at my watch I would destroy the balance and go falling away into eternity. Even at that stage of my journey I had learnt with surprise how easily solitude and loneliness are taken to mean the same thing, and they are not. Writing as I do now with the voyage completed, I know that the first and most common question asked me by all races, castes and creeds, was 'Isn't it terribly lonely?' And I came to see in many eyes—in the eyes of rich men surrounded by crowds; in the eyes of popular society leaders; in the eyes of beautiful women commanding wealth and adoration, or even of women with the security of a home and their own children; in eyes staring above this question in harbour bars, in native villages, in sumptuous apartments, in seething dock areas, in the packed hire-purchase homes of the Welfare State—and in these many eyes that question was too seldom interest, too often fear.

To this world-wide fear of loneliness, solitude is not necessarily the answer but that is where it may be found. The mere presence of others is not the answer, because some of the loneliest people I have ever seen have been immersed in raucous laughter six deep around a public bar.

And so days and nights passed as in a dream. I did all I could within the conditions imposed, and left the higher control of these to impose whatever limitations were intended, but the very fine line between fatalism and acceptance lies along a razor's edge. Thirst was often with me, because I made it a rule never to let the water reserve drop below two gallons. Hunger was also known when fish were elusive, and after two days mainly on jam and tea (there was plenty of both) I opened the last tin of sausages which was packed under the label of some obscure firm. They were disappointing: '…opened my last tin of sausages, which I bought as pork. They taste as much like pork as stray cats and dogs look like pigs.'

I bathed and soaped all over frequently in the icy squalls, because I feared boils or festering from my sweat-soaked skin. Even the sheet on my bunk was always sodden with sweat, and the twenty-four hours of each day were spent mostly in a pair of bathing Vs; I seldom bothered with oilskins.

And then on the night of 15 May I had a strange dream, and to this day I do not know whether I was awake or asleep. I awoke in my bunk and went above to find several people sitting in the cockpit, but obviously frightened by a heavy squall coming our way. I vaguely sensed they were near relatives, two of them women, but they were only fearful parts of their full personalities; their eyes and mouths were black holes, and their skin had an unearthly greenish tinge.

I knew it was my responsibility to reassure them but my nerves were on edge, so I first went forward along the side-deck and clung to the mast to collect myself. On return to the cockpit I sat on the starboard side of the helm, the only vacant seat, and opposite me on the port side was a stranger, seen dimly in the half-light enclosing the cockpit only. He was silent but strangely imposing, as if we all knew he was the leader. There was no noise of the sea, a complete silence.

I leant forward to the stranger, feeling ashamed, and asked, 'Do you think it will be all right?' but he did not answer. Sometime later he said, 'Read the Acts 14 to 19 daily for fourteen days.' And again later he said, 'You who ask so much, give a little more.'

It seemed to be impressed on me that this was not to be treated as a dream. We continued in the silence of the cockpit until he said something about a storm and waves covering the ship, '...but you and the captain will bring the ship into a cool place, tie up alongside, and all will walk ashore safely.' More followed, something about losing your life to save it and the laying on of hands, but I could not catch all that he said.

I wrote all this down next morning and looked up the Acts of the Apostles. In my Bible, 'To be read as literature', the Acts 14 to 19 are grouped within the Acts 13 to 21—did this explain 'the little more'? I read them for several days, finding no meaning in them, and then in a bout of bad weather missed a day and lost my belief in this strange dream.

Came unmistakable signs of the approaching monsoon, although the Pilot Book said it was not usually due until late June. In talking of the winds recorded it also said:

'The strength, however, varies considerably; the wind sometimes freshens to force six or seven, and gale force, or even forces nine and ten have been reached.' These last are hurricane forces.

Underlying the effect of local storms a persistent heavy swell developed from the south-west, and whereas before it had often been possible to sail round thunder clouds in my path they now became wide-based bastions, merging one with another to bar my way ahead. Sailing under these was like entering a darkened room. I held on to all sail in the light outer breezes and only reefed when the dark squalls reached out, sucking us into the gloom of torrential rain and the rumbling thunder overhead.

And then as *Sheila* tossed one night closely reefed between local storms, into my sleeping mind came the urgent need to get on deck, but each time I tried to rise from my bunk some strength forced me back. It could only have taken a split second to awaken fully, to find *Sheila* on her beam-ends and the scream of all hell above. I groped my way to the cockpit, where I stood on the side locker doors, as these were more level than the floor. It was pitch dark, heavy rain, and one consistent sheet of spray as the waves broke over the hull. *Sheila* was lifeless, pressed hard down by the tremendous pressures, and through the soles of my bare feet I could feel the strain the rigging was taking and only just holding. It was impossible to stand in that wind and I crouched clinging to the cockpit coaming.

Rightly or wrongly I decided to ease the pressures by running with the wind rather than by trying to get the sail in, and forced the helm hard up. Slowly *Sheila* turned down wind, I eased the main sheet and she staggered upright like a boxer rising from a knockout—and flew. I knew that it could not last without something going, but nor could I then leave the helm to get in sail—she would have swung into the wind and had the mast torn out of her. This was where I did need a crew.

The squall continued unabated but suddenly the helm became easier. The reefed mainsail, although only feet in front of me, was invisible in the blackness; the next flash of lightning starkly etched the tall mast and cross-trees against the darkness, the jutting spars streaming tatters of white canvas, and the gaff angled like a broken leg, broken near the jaws. Hell!

The squall passed about an hour later, when I hove to under storm canvas and went below shivering to get some clothes on. There was no point in worrying about the lost sail and broken gaff until daylight, apart from lashing down the latter securely.

At dawn the skies were fully overcast and were to remain so for the next few days. My last sights the day before had put me about 150 miles south-west of Pulo Bras, a small island on the northern tip of Sumatra, and it bore a light with a range of thirty miles. So from that last position I had to make a landfall on Pulo Bras, in appalling visibility, and then pass beyond into Malacca Strait. My only worry was what allowance to make for the ocean current, to guess when the south-westerly wind would change its direction—the difference between an adverse and a favourable current of one knot is fifty miles a day.

The loss of the mainsail was no real disaster, because the winds were either light, when I could use the big jib, trysail, the mizzen, another jib as a mizzen staysail, and either the genoa or spinnaker; or gale force or near it, when the normal storm canvas amply played its part. Anyway I was on a shipping lane and Penang was only 400 miles away.

On the evening of the second day a peaked smudge showed up fine on the starboard bow, exactly where Pulo Bras should have been. Rain blotted it out before dark, but I held my course, with reasonable hopes that its powerful light would show up between squalls before I hit it.

No light had shown by 10 p.m. and I was getting really worried, wondering whether the new Indonesian government had let it fall into disrepair. I could ensure not hitting Sumatra by easing my course to the north, but ten miles above Pulo Bras lay an un-lighted isolated rock, Rondo. Ten miles may seem enough for safety, reading this in front of the fire, but in the utter darkness enshrouding a small ship in half a gale, after seven hard weeks at sea, you begin to wonder if it is.

Then came the sound of breakers, and how the Almighty allows these to be heard above the tumult belabouring a rigged ship under those conditions, I have never understood.

I swung across wind to the north, still not sure that the island seen before dusk was not Pulo Bras. The next five hours were misery. The wind and rain had chilled me through, and after running into the sound of more breakers I tacked to the northwest, back towards the open ocean, only to pass more breakers to starboard. When the moon showed briefly through a break in the clouds I turned and sailed down the safety of its path, leaving the dark shape of the last island to port. Then the rain again drowned the moon, so I hove to and prayed for the dawn.

Daylight displayed a mass of islands—the Nicobars. I was off Car Nicobar, the northernmost, and about 120 miles away from where I should have been; whether this was due to a faulty sight three days before, or to a faulty estimation of the current, I shall never know.

The wind held about force six and although dead tired I sailed all that day, passing through the chain and to the east of Nancowry at noon and clearing Great Nicobar at dusk, a run of about 100 miles in fifteen hours. The wind died at midnight, and *Sheila* drifted helplessly with other flotsam into a long lane of debris where two conflicting tides met. I sat on deck till dawn, dead with fatigue, armed with the boat-hook and a torch. Several huge trees torn from some jungle coast were swirling and rolling in the tide, sometimes twisting towards *Sheila* until I fended them off. They could have smashed in her planking in the big swell that was running. At dawn the wind carried us away from the tide, but before leaving that ocean junk-yard I filled the cockpit with coconuts, which gave me food and drink until reaching Penang. Once clear of the rip I slept.

The monsoon loses its force inside Malacca Strait, and four more days passed before Pulo Langkawi off the Malayan coast showed up. I was sorely tempted to go a hundred miles north to Puket Island in Siam, where I had last been in 1946.

At that time I and thirty-six others had been flying in a RAF Dakota from Rangoon to Singapore, and at 10,000 feet above Puket Island the pilot discovered that someone had unfortunately forgotten to connect the reserve petrol tanks. During the peaceful glide towards the sea I presume we all did some serious thinking, but for several moments after contact at something like ninety-five miles per hour all thought was obliterated. An engine was torn from one wing and the aircraft stood on its nose before settling down to float like a ship. The onshore breeze soon washed us against the beach, when we hacked a hole in the side of the fuselage and walked ashore on the wing without even getting our feet wet.

For the next few days until we were unfortunately rescued, we played in the surf in the rubber dinghies and were entertained by the local Siamese Governor in a manner which we had previously read of as only being bestowed by one oriental potentate on another. The Siamese were wonderful hosts, and the first night the Governor took control we dined with him in a long shed, at a table running the length of it. At each place was a bottle of Siamese Black Cat whisky, and beside each chair was a dainty Siamese girl, whose duty it was to ensure that we lacked nothing.

After the board was removed small cocktail tables were placed down the centre of the room, while we sat with our Siamese hosts at other tables along the wall. There was music, and then followed my first introduction to Siamese dancing.

There it is the woman who asks the man to dance—a slim girl came gracefully over the floor towards me, and to my horror knelt at my feet in a humble gesture of supplication. Such a girl in the West would have had a queue of eager males fighting for a mere glance from her slanting eyes. However I soon got the hang of it, and thereafter if I was in the middle of an interesting conversation she waited at my feet until it was ended. This was civilisation!

She then rose and in time to the music glided round and round the small table, with body and arm movements which showed the decided influence of Hindu dancing. The man followed behind in the same manner, sometimes cutting a corner to attempt a capture, when with a twist of her body and a faint detached smile on her lips the girl escaped. It was a delightful version of the old chasing game.

This went on until well after dawn, broken only when we all trooped outside to the nearby beach, where we stepped out of our clothes and ran hand-in-hand into the warm surf. It was heaven.

On more formal occasions it was seen that these apparently oppressed females were treated with courtesy and wielded much influence, in spite of not having the vote so hardly won by their Western sisters.

On 21 May I wrote:

> Sailed all day under spinnaker and genoa and continued into the night as the breeze held. Sailed all night with a half-hour break at midnight for a mug of tea. Waves of feeling very sleepy, and nodding off at the helm.

Penang rose out of the haze about noon next day and the wind died. I had kept five gallons of petrol in reserve against the chance of this happening, because when the final destination is sighted it becomes an unreasonable but urgent need to attain it without delay. Soon after dark I anchored in three fathoms off the Swimming Club. The last tin of supplies on board, sardines, went into a steaming rice kedgeree, and after that I sat for a while on the foredeck with a pint mug of coffee. The hoot of cars came over the water and music from a harbour bar—and I felt wonderfully content that it should be so; and slinging a bright light in the rigging I slept.

24

*The Gurkha Officers' Club—the Home Guard—
the cost of living—the Malay girl*

As I look through my 1952 diary from which to write this chapter and relive the crowded six months spent in Malaya, I know that much must be left dormant in the past—there is too much to tell. And perhaps those small things are only important to me because I remember them so vividly; that first morning, after a deep hot bath holding more fresh water than *Sheila*'s tanks ever held, the sheer luxury of wearing a freshly laundered shirt following eight weeks of living like a naked heathen, the skin exposed to burning suns, icy rain, and the sharp flick of flying salt spray. Life was so very good—and fresh bread and butter every day.

And so too will I hurry over meeting port officials and the help they gave me, of securing *Sheila* at a safe mooring, and of my visit to the bank to find my worldly credit of $80 (about £10); and of there meeting Roger, who handed me a huge pile of mail. He became a good friend and I can still hear his sudden shout, 'Are you Hayter? Where the hell have you been?'

The next few days followed in searching for a job. With delight I found that the OC Area was Bill Simpson; we had travelled back from leave on the same ship years before and spent most of the time playing Liar Dice, which he always won. And then my hurried flight to Ipoh to fulfil a second object of my voyage. I stayed with Gemma and Peter whom I'd last seen in Aden, and the first morning I wandered round the Battalion Lines, to feel gladness in the sudden Gurkha grins that came with recognition. And that evening we went to the GOs'

(Gurkha Officers) Club—at first the introduction to new GOs and the semi-formal conversation, sitting round a long table, and the Gurkha orderly placing a glass by my side.

Aile katti mattri am dehrai soda-pani, I said as he produced the bottle of rum ('Only a little and plenty of soda-water'). And when all the glasses were filled I heard and said again, *Lo Sahib heru*. And in drinking I sincerely wished these men all the best.

Soon the years between had vanished and again we talked of mutual friends, bringing our respective news up to date, of Nepal and their homes, our present needs and past achievements. Sometimes as I turned aside to talk or listen I was aware of a cautious movement the other side, and turned back to find the orderly topping up my glass with neat rum.

Aile katti mattri Sahib, ('Only a little Sahib') he'd say as if knowing what was best for me—and continue pouring until I removed the glass in self-defence.

Finally in the early hours I dragged myself away from that circle of people, unsurpassed by any others I have ever known.

Lo Sahib heru, bato ko lagi— ('One for the road'). I rose reaching for my glass to find it brimming, the original soda-water long since lost in the surreptitious replenishings of neat rum. Courtesy and tradition stood on trial by ordeal.

I walked out of the mess as I had walked into it (I think). There was no such nonsense as, 'Are you all right, Sahib?' but several times on my way home I glanced behind to see a discreet shadow following—the GOs had sent an escort to ensure that I got home safely, and he would report to this effect before they slept.

Next day a telegram arrived from Penang to say that I was wanted for an interview, possible leading to a job in the Home Guard.

The object of the Home Guard then being raised was to arm and train villagers to look after themselves, thus releasing the regular security forces to concentrate more aggressively against the bandits. And so one morning found me standing in a glade of palms surrounded by thirty Indians, Malays, and Chinese villagers, about three of whom spoke a few words of English while I could throw a mere smattering of Malay among them.

During the week between courses, when one lot of students had left and before the next had arrived, I stayed alone in the camp and busied myself with the administration—pay sheets to check, undisbursed pay to be refunded, clothing and equipment ledgers to check, write-offs to be entered and accounted for, ration bills to pay and receipt, building contracts to arrange. Bamboo huts had by this time replaced tents and students numbered sixty.

I had received two letters from the other side, saying that I was perverting the people's minds and would pay the penalty if I did not leave. I kept these to myself as no one but myself could do anything about them, and most officials received them at some time. Some of course had proved worthy of attention, but if all had had the desired effect Malaya would have been in communist hands long before. But I did take precautions and was always armed, awake or asleep.

One day months later the police called and asked me to identify a body, a bandit who had been shot near a village in the huge area from which my students came. He was a Chinese and had attended one of my courses, apparently without benefiting greatly from my instruction.

The improved situation in Malaya, since I had last seen it in 1949, was basically due to the collection of the squatters into New Villages. These looked something like concentration camps, but within them the squatters could be protected and their produce denied to the bandits. As the bandit menace is neutralised the defensive measures will gradually be relaxed, and these villages will one day become the peaceful country towns of the future. But into this hopeful picture at that time came another factor—the cry for independence.

It amazed me to find that the loudest cry came from a Malay political leader, because the Malays had least to gain, being already ruled largely by their own Sultans, and everything to lose. Will their happy-go-lucky attitude to life ever withstand the persistent industry and business acumen of the Chinese? Unfortunately the press can be very misleading in these matters, because a bawling politician makes headline news while the ordinary citizen (if he bothers at all) is unheard. No Malay villager who spoke to me of independence viewed it with anything but alarm.

It seems to me that whereas before the War it was fashionable to say, 'Independence for the natives? Nonsense,' and the matter was closed, now the pendulum has swung to the other extreme and any politician, ambitious local, or malcontent prompted by 'the other side', can rise to his feet with a cry of 'Independence' and find enthusiastic support from the great democratic masses of the West. Perhaps the War left us adults more fatigued than we realise, because this same irresponsible attitude has also entered our homes. Whereas before the War we said to our growing children, 'Do as you're told,' we now say, 'Do as you like.'

Sometimes I toured the villages in my district, in the jeep, on foot, in canoes, to revisit past students and to assess what was being achieved; in some cases nothing, in others there was evidence to offer encouragement. In the Malay villages I was always met by the Head Man, and taken to his home while members of the Home Guard were rounded up to meet me.

We took off our shoes outside and climbed the steps into the house, as Malay houses are built on stilts three to six feet above the ground. It was always cool under the thatch roof, sitting on the floor on soft mats of woven grass, also used as beds. These people have an old-world courtesy which is very appealing, and often the children (after being hastily scrubbed and clothed) were brought in to give a shy *Salamat Tuan*. Chinese and Malay children are the most beautiful in the world.

In one village I was introduced to the oldest member of the Home Guard, a spry old Malay of 83 whose grinning, wrinkled face showed the years he carried but not their burden. He insisted I go to his home, where I also saw his three wives. One old crone had been the sweetheart of his youth, and two middle-aged women showed the fair wear and tear of motherhood—there were children everywhere.

I was told that my host was then looking for a fourth wife, the last of four allowed by his religion. On leaving I wished him well in his search, and expressed the hope that he would still find time to continue his support of the Home Guard. The old rogue roared his toothless delight, saying he'd do more than that; he'd increase it!

Another day, without warning, three armoured cars drove into the camp and out stepped General Templer. He asked a few questions and within five minutes knew what the camp produced and what prevented it from producing more—mainly accommodation and training facilities. These were noted down and he told me to contact Brigade next day and they would be made available. They were things I had been asking for over weeks past.

He then asked me how I happened to be in the Home Guard, and I told him, and for the next ten minutes we talked of trans-ocean single-handed sailing. He asked me to stay on another year, but I said my object lay elsewhere. General Templer was the most impressive inspecting officer I have ever met, putting his questions not with the object of finding where you went wrong, but on what you needed to do better—and then seeing you got it.

At this time there was a growing outcry from Government officials and employees—Army, Police and Civil Servants—for an increase in wages to meet the rising cost of living. In Malaya such things as servants, a car (at my seniority and above), and membership of a club had stepped outside the sphere of luxuries and were regarded as necessities. A Government Committee was later appointed to enquire into the matter and on its recommendation increases were sanctioned.

During this period I joined no club, drank only occasionally and not as a routine, kept no servants, and on this inadequate pay saved £100 a month clear. I don't recommend that way of life, because if you don't conform to the standards of the tribe you become an outcast; but since then whenever I hear of mass demands for higher wages I wonder who draws the dividing line between luxuries and necessities.

There was some social relaxation during this period, because I cannot do without it when it is available, just as I can only give up cigarettes when I can't get them. The Government-supplied jeep was for duty purposes only, but when I drove from camp into Butterworth and crossed on the ferry to attend a conference in Penang, I usually combined this with some amusement, or a flying visit to *Sheila* where she was moored about five miles south of Penang. She was in full commission, and it was sometimes possible to take friends sailing on a Sunday. Sometimes I spent the night at Roger's house, or he took me to the Club, and sometimes I went to see a pretty Malay girl with a beautiful figure.

In my own circles, among my own kind and class in the hotels and clubs, I felt I could seldom make any real contact, and this was particularly so immediately after arrival from the sea when I needed it most. I could not get the party spirit, and a strangeness seemed to spread subtly outwards from me to dampen the ardour of those who did.

I danced with this Malay girl several times. She had a quiet dignity, and laughed at my Malay conversation until she spoke in much better English. One evening, and others later, I suggested we drive out to a beach and she agreed—we took a rug and a couple of bottles of beer, a packet of sandwiches or some strange food she bought from a Chinese street stall. The tropical nights were warm, and often we swam in the dark water milky-white with phosphorescence. The hours drifted past as we talked of many things—she had no education, but she had thoughtful views to express or questions to ask on almost anything.

Sometimes there were long silences between us, when I became lost in thought about my book or in some problem connected with the camp. She sat in silence beside me, or rose without a word and wandered off into the starlight to sit on a bank or against a palm, lost in thoughts of her own. These silences brought no uneasiness nor the desire to break them, and they usually ended by a mutual understanding which brought no discord to the other's thoughts.

She came to sit beside me and asked, 'You go with other girls?'

She had no recognised claim on me, yet the question came naturally from the sympathy between us.

'No,' I said, 'I don't go with other girls.'

Slowly she turned, and putting her arms around my neck she laid her cheek against mine and stayed like that a long time, very quietly.

This is perhaps the greatest gift that one human being may receive from another; it can neither be given nor sought with deliberate intent, but if the way is opened giving and receiving become one by which to enter a great stillness. It goes beyond any difference of race, colour, creed, social status or academic background, this unity of some spark which lies within us all, this state of wonder and beauty by which we touch eternity.

It goes too beyond sexual desire, although there is a subtle connection between the two and one is easily mistaken for the other, always to cause great hurt.

25

*Malacca Strait—night entry into Singapore—
social engagements—departure*

The voyage of about 350 miles down Malacca Strait to Singapore took a fortnight. Doldrum conditions prevailed, calms frequent, winds unreliable, and although navigation was sometimes worrying during darkness because of numerous coastal reefs and shoals, my greatest concern was the Sumatra Squalls common at that time of the year, and usually arriving at night between 10 p.m. and 3 a.m. They came out of the west in a thick black line of cloud which ate up the stars, bringing terrific thunder and lightning and a blinding downpour. There were no outriding squalls preceding these storms, but the leading edge of the wind could be seen on the darkest night where it whipped the black water into phosphorescence, and then struck the ship with solid impact. I had to beware that they did not drive me on to the shallow lee shore.

The coast was shallow, the five fathom line in places as far as fifteen miles or more offshore, and so the long barriers of sturdy fishing stakes were also a menace at night. Quite often a bamboo house was built on the outer corner of these, high above the swift tidal water, and it must have been terrifying inside during a Sumatra. In daylight I often sailed within feet of these houses, in which lived several Chinese families, with fluttering hens perched on the roof and naked babies crawling about on the open verandahs. I wondered how often Chinese mothers, busy with their housework, asked, 'Has anyone seen baby lately?'

I rounded the southern tip of Malaya under bare poles, in the smothering rain and gale-force wind of a Sumatra, and during the following day passed Sultan Shoal and Raffles Light, which mark the limits of a huge area of islands and reefs projecting south into Main Strait from Singapore Island. After dark, with all sail set and the engine at full throttle, I could not drive *Sheila* against the strong south-westerly current to get into Singapore harbour, and so I sought an anchorage near two small islands called the Sisters, until the tide should turn.

It was pitch dark, but the vague outlines of the two islands showed against the sky. The chart showed that the sea-bed rose steeply from deep water to a small shoal projecting south from the islands. It was on this I decided to anchor, and planned to avoid the neighbouring reefs by keeping within the limits of two compass bearings from the islands.

Owing to the difficulty of judging distance off at night I approached the shoal on soundings with the lead, to find water shallow enough to hold my anchor but not to proceed so far as to hit the rising shoal itself. *Sheila* will not hold her course under power as she will under sail, so leaving the engine running in neutral I let *Sheila* sail herself through the darkness towards the dim island shapes, while I took soundings from the side-deck. The tide carried her off the line of the shoal before soundings found it, so using power and sail I circled and cautiously ran in again, to find the end of the shoal before being swept off it a second time. I ran in a third time, anchored in four fathoms, and slept for four hours.

At 2 a.m. any change of the local tide had not affected the prevailing current, but a thick line of black cloud gave warning of a Sumatra Squall which struck as I broke the anchor clear, and *Sheila* flew before it under jib and mizzen, her bow sucked and shouldered off-course by the swirling current.

One of the most difficult feats of navigation under sail, in my experience, is to enter at night a huge harbour packed with shipping, in a heavy wind and blinding rain. Other than shipping anchored in the stream the chart showed wrecks, shoals, unlighted buoys, all of which had to be avoided, and the only way to ensure doing so was by taking continuous compass bearings from recognised lights and buoys. These only showed spasmodically through gaps in the rain, and were then confused with the lights of shipping and the myriad winking coloured lights of the city beyond. I did not dare leave the helm travelling at that speed, and transferring compass bearings accurately on to a chart with a hand-torch, in a drenched and wind-ravaged cockpit, became a juggling act in itself. Several times lights appeared mistily almost overhead, and the great hull of a ship towered blackly in time for *Sheila* to foam past her side, or slide clear past the straining anchor chain.

Memories of the harbour from other days made the entry easier than it might have been, and after picking up the green flashing light on the southern end of the breakwater it was comparatively simple to find the Yacht Club. I took *Sheila* in behind the half-submerged hulk of a war-shattered wreck and anchored, much to my relief.

The Royal Singapore Yacht Club is an imposing building standing in its own grounds, and the members gave me a friendly welcome and all offers of help. I later changed my berth only because the anchorage is so exposed to the full force of Sumatra Squalls; the *California* had anchored there and very nearly been driven ashore. So after a few days with an introduction from Doc Mooney RN, (who had known *Sheila* in England before I bought her) I moved round to the Naval base sailing club, almost as far as the causeway in Johore Strait which separates the Island of Singapore from the mainland. There *Sheila* was moored between two safe buoys and I lived on board.

At this time, the end of January 1953, it was my intention to complete *Sheila*'s refit and leave as soon as possible for the Java Sea before the North-west Monsoon ended. From many enquiries made since Colombo I found that the unsettled conditions in Indonesia, and the antipathetic attitude to Europeans, made it unwise to call there. I planned to sail therefore straight through to Dilly in Portuguese Timor, with a surreptitious call at the Island of Komodo in order to photograph some of the huge lizards, up to fifteen feet long, which still live there in great numbers. My route thereafter was planned through Torres Strait (between New Guinea and Australia), and then down the Great Barrier Reef to Sydney.

The completion of *Sheila*'s refit and the repayment of loans which had helped me to clear India came to £400. This left me £200, but before transferring this to Timor and Sydney to meet needs in the future,

I spent it. There were certain difficulties which delayed me, but they need not have taken three months; perhaps the real reason was that for the first time in three hard worrying years I was at peace mentally and had money to spend. Also, nearly all the club members were in one or other of the services, there were old friends in the First Battalion of my regiment stationed in Singapore, and I felt that the friendships renewed and formed were of more value than submission to the demands of Time. Nor did I ever have any regrets in spite of what followed. My search lay in the sea and the thoughts it induced, but the experience of human relationships ashore are needed to interpret those thoughts from idealistic theory into a code of practical living.

One cause of my delay was undoubtedly Peter Hamilton whose yacht *Speedwell* was moored on the same string of buoys as *Sheila*, and we were refitting at the same time. He was preparing for a voyage via the Cape of Good Hope to England (where he arrived in due course and then sailed to Canada. I have heard he is now getting married and intends to continue his journey *en famille* into the Pacific).

The heat of course was intense, and about 11 a.m. on many days the conversation went something like this (names could be interchanged):

'Are you making satisfactory progress, Adrian?'

'Yes, pretty good, Peter. Slowing up a bit.'

'It is hot, isn't it?'

'You know, Peter,' I'd say, 'these clever psychologist blokes reckon that if a worker takes half an hour off from a job for relaxation and refreshment, he returns with such increased vigour that he actually does more work finally than if he had taken no time off.'

Here Peter would stop doing whatever he was doing and give his full attention.

'Adrian, do you think we uncouth sailors should ignore such wisdom?'

'I don't think one would do us any harm, Peter.' And so we would row the few yards to the Red House (the Clubhouse) for a helpful glass of cold lager.

Unfortunately there we nearly always found two or three pleasant people, and quite often my very special girl friend. Victoria was, I think, about four, and already knowing the power she wielded over susceptible males she used it ruthlessly. If she climbed suddenly off my lap and went to whisper in her Mother's ear it meant that I had to go back to lunch with them, and the rest of the day was no longer at my disposal. Bill and Joan, her parents, became my good friends.

Bill overhauled *Sheila*'s lighting system and electrical parts, doing a much better job than the many professionals who had previously charged me great sums for achieving little. Sometimes in the afternoons we sailed over to a tiny beach on one of the lovely small islands in the Strait to bathe and grill chops, and return for Victoria's bedtime.

There were five or six young girls who lived in their own mess nearby—school teachers, nurses, welfare workers, or those of similar occupations organised by a Home Government for the betterment of Malaya or the world generally. And in almost every port sooner or later I met a girl called Sheila who of course had special privileges, meaning that I could usually induce her to scrub out the galley and impart a feminine touch to those tasks for which my sex did not befit me.

Eventually the club Committee went to much trouble to arrange a farewell party to Peter's *Speedwell* and my *Sheila*; we had agreed to sail on the same date but when the day arrived I was still not ready. And so Peter departed on his voyage but *Sheila* stayed serenely at her mooring.

Such a failure on my part perhaps gave an air of anti-climax to the gathering, and I have often felt the disappointment I have given to others of not fulfilling the usual picture of what a single-handed sailor should be—taciturn, ruthlessly efficient, and showing no glimmer of emotion except at talk of the sea and ships. Yet you are also expected to drink heavily and be very lustful.

One person had said to me in a previous port, 'You know, you're very disappointing to meet. What you've done is unusual and I expected to meet the hell of a tough; but you're just an ordinary person such as one meets every day.'

Fortunately by that time I had learnt to recognise, after a few minutes' conversation, anyone who has read that famous best seller *How to*

Win Friends and Influence People, so I accepted this candour with the friendly spirit in which it was given.

An efficient Naval officer had also said, 'I would love to do what you're doing but I wouldn't sail alone. I'd be afraid of getting slack and inefficient.'

However inefficient my preparations may have appeared to a casual observer in port, they were as thorough as finances permitted—even my expensive relaxation in Singapore was consciously indulged, because by saving too ruthlessly you can kill your efficiency as certainly as by squandering too freely, which I never had enough to do anyway. But at sea I imposed a discipline far more severe than that prevailing in the services, simply because it was more necessary; to the sea a reason for failure is never an excuse—an error through sheer carelessness or lack of knowledge will have the same result.

I eventually said goodbye to friends in my own time as I hate being seen off; many will know that dreadful feeling of a fond parting at a railway station when the train simply will not go. And so finally I sailed in mid-afternoon on 13 April, 1953, in a blinding squall; the hardest farewell had been the last, to little Victoria. I have been bruised by many farewells, but a child lives in eternities and when these are suddenly ended it seems an injustice beyond their understanding. They give so generously of that which we all need so much, and by departing for reasons which are meaningless to them we seem to reject it.

26

*Planning problems—a sea of islands—the danger of rum—
under armed arrest—mine-field—arrive Surabaya—the Coronation Ball*

This voyage from Singapore to Portuguese Timor was quite different from the others. Those had traversed empty miles of open ocean, whereas this for the most part followed a winding channel through a maze of hundreds of islands, reefs, atolls and shoals. In the open ocean away from land and shipping, sleep had been no great problem, but in these confined island waters I had to seek safety in which to sleep, and before sailing I spent detailed hours over the charts selecting possible anchorages for use should the need arise. Remembering the dread force of Sumatra Squalls I chose havens protected from the west, the direction from which those squalls came.

April was the transition period between monsoons, and the Pilot Book stated, '...the winds are light and variable, and squalls and thunderstorms rather frequent.' So I knew I could not expect a fast passage. If the south-east monsoon (as it is in that part of the world, being an extension of the south-east trades which blow in the Indian Ocean below Java) hampered me too much on my way to Timor, I planned to lie up for six months in the island of Komodo. It was uninhabited, had a good anchorage, was teeming with deer, pig, and wild horses; and in that time I should be able to get some really good photos of the famous Dragon. The Pilot Book's description of these brutes promised excitement: 'A peculiar creature, sometimes called the 'Komodo Dragon' owing to its resemblance to that legendary monster, is found in the forests of Komodo… these beasts possess colossal strength, and attack the numerous wild horses on the island, and sometimes even men.'

Another hazard to be faced on this voyage had nothing to do with the sea, but with international politics. The Indonesians were very suspicious of any curiosity about what was going on in that part of the world, as witnessed by the lack of news allowed out of it. People who knew told me that the chances of being flung into gaol or even being shot out of hand were not small; such incidents of course make interesting reading in a travel book, but my object was to arrive in New Zealand sufficiently undamaged to write one.

For these political reasons all the anchorages I chose were at islands which the chart showed to be without water, and therefore unlikely to be inhabited. (Komodo had been reported uninhabited, but I knew it would have water because of the numerous animals living there.)

The passage southwards down Rhio Strait was slow, drifting through frequent calms, beating against southerly winds, and hastily saving the sails under the onslaught of the thunder squalls. These had to be watched and anticipated and courses shaped accordingly, because when stripped of all sail *Sheila* was at the mercy of the squall and the strong currents which snaked among the islands. There was little enough lee-room in the narrow channels for peace of mind, and it was often worrying waiting for the wind to ease enough to allow me to get on some sail and regain control, to clear *Sheila* away from a reef or island. At night these conditions were particularly worrying, and I often felt that the success of the very accurate judgment required was more the result of luck than of my own ability. And that is no way in which to navigate.

I have come to believe that if you do all in your power to meet a difficult situation which has to be faced, and spare yourself not at all, then luck will attend you and give the extra that is needed over and above your human capabilities; but if you begin to exploit this, to take inessential risks purely for your own convenience and then expect luck to attend you as your due, you will fall—and heavily.

The conditions prevailing in the narrows of the Rhio Archipelago were very similar to those to be expected within the Great Barrier Reef north of Australia—strong tides, heavy squalls and narrow coral-congested passages. To meet such conditions single-handed with merely a weak auxiliary motor seemed to me to be tempting Providence, and I wondered again whether it would not be wiser to sail home south of Australia instead of via the Barrier Reef.

Sometimes during that first week of the voyage there was no apparent outlet to the way beyond, when the hundreds of islands interweaved and overlapped to fill the horizon. Most of them were low-lying and palm-covered, their guardian coral hidden beneath the dancing waves; some were large with high, jungle-clad peaks, and when the low ground of the more distant of these was hidden beyond the horizon, leaving only the hilltops to show isolated, these too looked like small islands and did much to confuse navigation. Others were only strings of palms on a stretch of coral sand, mere feet above the water.

It would have been very easy to have taken the wrong turning in some places, and continual compass bearings were necessary. After my sextant the hand-bearing compass was the most useful navigational instrument on board. By bearings from recognised islands abreast and astern I plotted my position on the chart, and then from that position I laid off bearings on the chart to the islands ahead, and then recognised them by viewing them over the converted compass bearing. This was the only way of being sure of following the correct channel, and also of avoiding various underwater dangers which projected into it.

I slept for two hours after dawn on the sixth day out, following an unbroken twenty hours at the helm, and for the remainder of the day headed for an island where I determined to anchor and rest. This island, only two miles off the equator, was shaped like a crescent with the open side sheltered by another smaller island. The two arms of the crescent reached out to this latter in coral reefs, but on one side the chart showed a narrow entrance. I planned to get through this on bearings from the hills, then go to the head of the bay between the two protecting arms and anchor in four fathoms.

I was still five miles off when darkness fell, but the island showed up in the brilliant starlight. It is dangerous entering strange coral reefs at night, but increasing fatigue and the denial of a good night's sleep can become more dangerous than the entry itself. I moved down along the coast until starlit water showed the gap between the projecting arm and the smaller island, and when the dark outline of the hill-top came on to the bearing taken from the chart, I put the helm over and slowly entered the reefs; when the point of land was aft of the beam I knew that I had cleared into the comparatively open water inside, and only then realised that my mouth was quite dry from the anxious moments passed.

Sheila eased along the curving shore inside the lagoon, the faint breeze just keeping her sails asleep, and when we entered the area of water darkened by the reflection of the hill behind, making it impossible to judge the distance off the beach, I moved further in on soundings with the lead until finding four fathoms, when I dropped the anchor.

The sails came down with the lovely sound of dry ropes running easily in their blocks, and after stowing the sails I went below to light the lamp and prepare supper. While some rice cooked to eat with the remains of a bully-beef stew, I sat on deck with a tot of rum listening to the growl of the surf on the outer reef, and the audible stillness of a lagoon at night. After a hot meal and a big cup of coffee, I slept.

Dawn brought a heavy squall, which came racing over the reef between the two islands, heeling *Sheila* far over and forcing her to the limit of her chain. The one anchor in the soft bottom mud could not hold her, and I frantically unshipped the dinghy in the driving rain and laid out the second anchor, which halted *Sheila* twenty yards from the coral fringing the beach.

If that squall had come at night, being very tired and doped with rum I might not have awakened in time, and *Sheila's* bones could now be on that coral. It is madness to lie to only one anchor in those waters, and madness to accept the comfort of alcohol except in a really 100 per cent secure anchorage—something I have never found.

This is why I never touched rum in the forward areas during the War on the few occasions when it was available, because you only borrow on the future; when the repayment is demanded you are more useless than you would have been bearing the fatigue so momentarily banished. These comforts are only sensible when you know that no further immediate demands are to be made against you—and at war or at sea this is something you do not know. A bottle of benzedrine carried on board for extreme emergencies remained unopened during the whole voyage, not because there was never an extreme emergency, but because I could never be sure that it would not be followed by another.

Early the next morning I crossed the Equator, and it was a thrill to know that at least I'd arrived in the same hemisphere as my home. Thereafter it was slow progress over the next two weeks creeping through Gasper Strait between the large islands of Banka and Billiton on fickle winds, delayed by long stifling calms, and cursed by the violent thunder squalls. Owing to the intense heat there was a heavier drain on my fresh water supplies than usual, and I decided to call at an island to fill my tanks before continuing on to Timor.

Nine days later the high peak of Bawean showed up, and I closed its coast two days after just before dark. The small scale chart did not give enough detail on which to enter the reefs in that light, and so I laid off and on all night and headed inshore soon after dawn, passing through a fleet of fishing canoes but sheering aside from those who tried to board not sure whether their excited yells were hostile or friendly. The entrance through the reef was hard to find and difficult to negotiate, and the whole village was aroused by the time I dropped anchor.

Several canoes put out, and the first locals on board were eight armed soldiers. The commander explained that the ship was his and that I was under arrest, and having detailed two soldiers to stay on board he took me ashore to face the local Commandant.

The judgment took place on the verandah of the police station, a whitewashed adobe and bamboo building almost hidden in the thick fringe of banana palms bordering a cleared square. The whole village turned out to watch—I was told that I was the first orang puteh (white man) to land on Bawean since the War, and to children of ten years of age and under I was unique. The discussion was polite but lengthy, partly because the crowd increased the local commander's sense of his own importance, and partly, I suspected, to give the two soldiers time to search *Sheila* from stem to stern. It had not been possible to hide my rifle and ammunition and the only other valuables were my cameras.

These I had buried in two four-pound tins of rice before entering the reef, and doubted if they would be found.

I was allowed back to *Sheila* and immediately began filling the water tanks. I only had a four-gallon container with which to ferry nearly forty gallons, and as *Sheila* was about half a mile offshore and the water-point (an unhealthy-looking well) was several hundred yards inland, I barely completed the job before dusk, and being tired went to bed early. The escort had returned to the shore (the exit through the reef was too dangerous to attempt in the dark), but I was awakened about 9 p.m. when a heavy canoe bumped alongside, and the military commander came on board.

He explained that he must take my rifle and ammunition, my ship's papers and passport, for despatch to the authorities in Surabaya (informing me to my surprise that these orders had come by radio). Officials in all the lesser countries always demand these items, but according to International Law they have no right whatsoever to remove them from the ship. In this case it was pointless to refuse, and anyway I would not hesitate to sail without them should the opportunity occur. I was also beginning to wonder whether such an escape would be sensible, remembering the radio link to Surabaya. They could send out a plane or patrol boats which would probably pick me up in no time, and they might not hold their fire.

That night I moved *Sheila* nearer the exit through the reef, telling the escort in the morning that she had dragged her anchors. At noon when the sun was high over the coral I went aloft to work on the rigging, but more to note the lighter-coloured water of the passage and try to work out bearings which would get me through it in the dark—bearings on to the high peaked hill which would show up as skyline during a starlit night.

In the afternoon I called on the local Governor, who was a very pleasant person with the charm and manners of the usual un-warped Malay. He occupied a house which had probably been Dutch before the war; anyway, it had a long bath and a shower which were made available to me. In the days that followed I spent much time with this family, teaching my host English (which is now the second language of Indonesia) and learning Malay from him. He replenished my supplies of cigarettes, and in return I gave him enough 2in. manilla rope to restrain his tame buffalo from ravaging the vegetable garden.

In the mornings the village market was a riot of colour and chatter, and I often wandered through it because the ordinary people were friendly and amusing—the local civilian Governor, for instance, had been so, although he held no status at all when even a military lieutenant was around. It is only the armed people in strange countries who are bumptious and unpleasant, and this is why I now suspect the inner integrity of any country whose police are armed.

The atmosphere of the island began to grip me, and I slowly realised that my prison was one of the most heavenly places I'd ever visited. Even the soldiers were not too unpleasant, though the Lieutenant in charge was always rude and insolent, and my right foot itched whenever he approached. Each morning my military escort came on board at dawn, with the news that orders for my release would come from Surabaya tomorrow; but they left at dark.

However I could not delay much longer, simply because I was living on the ship's supplies and would soon have insufficient left to take me on to Komodo (whereafter I could live off the land). I did not want to leave without my rifle, but I could live on Komodo without it as turtles there are plentiful and are easily caught by hand. By this time I had completed a home-made chart of the exit through the reef, had carefully checked bearings with the courses of dhows entering and leaving, and had noted the tide and wind prevailing over the hours after dark.

On 20 May, after five days in Bawean, I decided to sail that night, hoping that the authorities in Surabaya would be glad to be rid of me, rather than consider me important enough to send forces to collect me. There was little surge that day outside, and I planned to leave with the offshore breeze under the jib and mizzen only, so as to have full steerage but not enough speed to do great damage if the reef did get in the way, when I would use the engine to get me off. With the silence of sail there was every chance that my departure would not be known until the following morning.

An hour before dark I knew that I had left it too late. A long grey motor vessel entered the reef and anchored just ahead of me, opposite the exit through the reefs. She carried another section of troops, all heavily armed with tommy guns, rifles, swords, and several had our 36-grenades clipped to their belts. Escape under sail was certainly out of the question for that night anyway, and so I went to sleep wondering what the morrow would bring.

Soon after dawn a heavy canoe bumped alongside and the Lieutenant, the Governor, the Military Commandant from the other side of the island, a Naval officer, together with six heavily armed troops clambered on board without courtesy. I resent anyone coming on board *Sheila* without my invitation, because she is my home and any home is sacred ground.

The Lieutenant introduced the Commandant, who was only a shabby Captain, but it was this latter who said in Malay, 'You are ordered to go to Surabaya.'

This I flatly refused to do, explaining that I had called at their country merely for water, and that they had no right whatsoever to interfere with a British ship (neither of which in these days meant a thing!)

'Then we will tow you,' and he signalled the Navy vessel, which immediately began to get in its anchor. It was more the frustration of the situation than the thought of going to Surabaya that aroused the anger; I took the Captain by the arm, pointed to his canoe, and in no uncertain English told the whole lot of lousy bastards to get off my ship—it was too difficult being angry in my limited Malay, but anger carries its meaning beyond the words of any language. Immediately one of the soldiers close to me in the confined cockpit dug the muzzle of his tommy-gun into my ribs, and calmly glancing down changed the catch from safety to automatic. There was no hate or anger in his eyes but I could see that he had killed before. I suddenly received the conviction that this was no story-book, that if I did anything silly I was going to get myself killed—and no one in the outer world would probably ever know a thing about it. So I let go the Captain's arm and sat down on the cockpit coaming.

Outside, the first of the new monsoon had arrived, stirring up a rough sea, through which the 100-mile tow at a speed of at least ten knots would have strained *Sheila* badly and done great damage. This I explained to the Navy skipper, who agreed and supported my plea to the Military Commandant, which was that I be allowed to proceed independently to Surabaya, having given my word to do so. This was accepted and the Naval launch left immediately.

Sheila was all ready for sea and I sailed within half an hour. As I drew clear of the reef and headed west to round the island before taking the southerly course to Surabaya, three big dhows followed. One took station a mile up wind, another a mile down wind, and the third a mile astern. The huge lateen sails gave them the same speed as *Sheila*, but she could beat them in heavier winds or in lighter with the spinnaker or genoa set.

Then followed a wonderful, exhilarating experience I will never forget. It was good to be at sea again, this arrest by a foreign power was full of interest and I had no idea how it was to end, and the monsoon was blowing strong on the beam, raising big seas and driving *Sheila* hard across them. The escorting dhows gave a sense of rivalry, and looked magnificent as they rose to the crests under the huge spread of sail, sank almost out of sight in the troughs, but always keeping station. I held on to all sail at dark, because I was going to need all the speed *Sheila* could give.

The only thing that did worry me was the fact that no one on my side had any idea where I was, except that I was somewhere between Singapore and Timor. From information previously received it was quite likely that the Indonesians would throw me into gaol, without notifying the British Consul or allowing me to contact him. For an hour after dark I used a light irregularly, so that the escorting dhows could see I was still on course but would not worry about periods of blackout.

At 9 p.m. I doused the light and turned hard down-wind, passing astern of the dhow on that station and across the bow of the one astern, without seeing either of them in the darkness. Then I turned north towards the shipping lane, with the idea of contacting a ship to take a message to the authorities in Singapore.

No ship passed that night, and at dawn I turned south again headed for Surabaya under reefed sails to lessen my speed. I thought that when the dhows arrived without me the Indonesians might send out a plane or patrol boats, and assuming I had broken parole, open fire. This was less likely if they found me heading towards Surabaya.

At dark I again turned back towards the shipping lane, and sometime before midnight flashed up a ship. She answered, but owing to her distance away and the intervening waves I could not be certain that she had received my message, so I sent SOS and asked her to speak with me. This was the only SOS sent during the whole voyage to another ship, and I gratefully saw her lights turn as she altered course towards me. Soon the two ships were lying close in the darkness, the laden tanker with her decks awash, her bows flaring high above to make *Sheila* feel very small indeed.

Clinging to the rigging I shouted my story to the mate, who strangely happened to be a New Zealander, and who promised to pass my message to the Foreign Office in Singapore. With that problem off my mind I turned south once again, headed for Surabaya to fulfil the promise I had given.

There were strong winds and currents from the east on the Java coast, and having lost easting I had to beat backwards and forwards for three days and nights to make the entrance to Madura Strait, which leads to Surabaya. My chart showed a fifteen-mile buoyed channel leading into this Strait, and this I could not understand. There was plenty of water even for ocean-going ships, no off-lying shoals marked, and the only dangers were a few wrecks which could be avoided by bearings. As it meant beating another ten miles against wind and current to make the entrance to the channel I ignored it and made Madura Strait direct.

I had no large-scale chart of the twenty-mile tortuous Madura channel, and being the guest of the Indonesian government I decided to enjoy the luxury of a pilot, and at dawn signalled the light ship which housed them. Owing to the rough conditions, he had great difficulty boarding me from the pilot launch, which was handled with great skill, and I fully expected him to refuse when my mainsail suddenly tore right across. However he got on board eventually, I put *Sheila* under jib, mizzen, and engine, while the pilot took the helm and I stitched the mainsail. He was an Ambonese, a people whom I grew to like and respect because they looked something like and were similar in characteristics to Gurkhas. He kept roaring with laughter and shouting, *Bagus, bagus* ('Good, good'). With the mainsail up we later had a beautiful sail in the afternoon.

I had been ordered to anchor at the police pier on arrival in Surabaya, and on my way to that point recognised what was undoubtedly a Dutch yacht club, so hastily turned *Sheila* aside, dropped an anchor, flung the dinghy over the side, and shortly after was on the telephone to the British Consul. Within half an hour I was sitting with a young couple, the Vice-Consul and his attractive wife, discussing the possibility of attending the Coronation Ball to be held at the Consulate.

At the first opportunity I asked the harbourmaster the meaning of the marked channel leading through deep water into Madura Strait.

'Good heavens,' he said in English, 'show me the way you came.' So I showed him my course from Bawean, and the area off the coast where I had tacked backwards and forwards against wind and tide for three days.

'All this area here,' he said, indicating either side of the channel and including where I had tacked back and forth, 'is a magnetic minefield. It was laid during the War and only that marked channel has been cleared for the use of shipping.' And *Sheila* has about three and a half tons of iron on her keel!

I felt like Cinderella three nights later, dressed in my sharkskin dinner jacket and reclining in a large consular car while a smart chauffeur drove me to the ball. I was allowed out on the Consul's guarantee of my good behaviour. And so I drank the health of our new Queen.

During the evening I danced several times with the very attractive young wife of one of the senior officials who had ordered me to Surabaya. She wore a beautiful evening gown, danced like a dream, spoke fluent English, and had a lively sense of humour. She asked about my voyage and listened with amusement to my account of the closing stages.

'And what,' she asked, 'now impresses you most in Surabaya?'

'This,' I said.

'But you've been to a ball before?'

'Yes,' I admitted, 'but I've never before been placed under armed arrest by a foreign power; nor released briefly to attend a Coronation Ball; nor enjoyed dancing so much with the wife of a tormentor.'

'And what did you expect?' she asked. 'A lot of barbarians?'

27

Dutch hospitality—a spoilt child—political thoughts—departure

At the ball I also met a young Dutch couple, who kindly asked me to stay a few days as a rest from living on board. I gratefully accepted, because the harbour was dirty and I was confined to the fringe, consisting of warehouses which always depress me for some reason, and the usual harbour honky-tonks. I had long since passed the stage when these held any fascination for me, because they are the same in any port in the world—the drab curtains and untidy bar, the haphazard music, the sailors of varied nationalities (the British and American usually young and acting tough, giving the impression more of washed-out Teddy boys than iron men of the sea), the tired shoddy women, and the inevitable male perverts.

So when Kitty and Fien arrived two days later saying that all formalities were arranged, I threw some clothing into a suitcase, locked *Sheila*, and departed to a pleasant home—to sleep between clean sheets, bathe daily morning and evening, linger through good meals, and meet an ever-widening range of new people.

Time passed quickly, and by this date I should have been in Komodo. Then suddenly the Indonesian authorities cleared me from arrest and issued the order that I must sail in three days, whether I was ready to do so or not. By this time the full force of the monsoon had developed and any voyage to the east was out of the question.

There were about ten days' supplies left in *Sheila*'s lockers and the only answer that I could think of was to return to Singapore, which with the monsoon astern would only take about ten days. There I could get back my old job in the Home Guard, but would have to sign on for at least a year and then make this same trip all over again. This was turning back, and I have a superstition about that—which, unfortunately for her, Lot's wife had not.

Fien asked if I wanted to return to Singapore, and I told him that it was no longer a matter of what I wanted or didn't want— it was entirely my own fault for having weakened and taken a holiday there instead of sailing earlier; I had no complaints.

'But why not stay here with us?' he asked, as if it was the most usual thing in the world to invite a stranger into his home for the coming six months. Anyway, that is what eventually happened—for the next eight months. I don't know how he arranged this with the authorities.

Kitty's and Fien's home became my own and I lacked nothing. They took me out wherever they went, unless I was busy writing; their friends accepted me in the same way; Fien told me to take his car whenever he was not using it; the servants looked after me as an honoured guest; twice a week on entering my room I found a carton of cigarettes lying on my bed, and two or three hundred rupees in cash.

'Fien, I can't take this—I don't know when I shall ever be able to pay you back.'

'Never mind, some day, when you get to New Zealand.'

'But I may never get to New Zealand.'

'Never mind,' said Fien. 'Some day perhaps; but now you must have some money to spend. It is nothing.'

How is it possible to receive so much kindness and generosity from people who were initially strangers of a foreign race? There was no reason why they should have bothered about me. I grew to love those two, not for all they gave me but for what they were; I shall never forget them.

But first there was a threshold to cross before the harmony of that home was mine, and that threshold was the young son of five. He was an only child, spoilt, and he resented this entry of a stranger into his domain of security and affection. In his parents' presence we accepted each other with outward civility, but an antipathy grew between us which we both felt.

Everyone slept during the heat of the afternoon, but sometimes I sat on the back verandah to read or write. And one afternoon it was here that the child found me.

He approached, staring steadily at me, and stopped just out of reach—then spat.

'You little bastard,' I said, and in Malay (which he knew from his amah—I knew no Dutch) told him not to do that again. This he promptly did with increased volume and accuracy. I told him if he did so again I would smack him hard—he of course knew very well that I would not do so for fear of his fond mother's reaction; and I knew that if he called my bluff he would thereafter make my life hell and we would grow to detest each other deeply.

Again he approached, looking me steadily in the eye, and spat. I grabbed, and before he knew what was happening he was experiencing his first hard whacking with extreme distaste. Then he ran screaming to his mother, who came out half an hour later.

'I see you and Era are beginning to settle down together,' she said.

'Yes,' I said. 'I do hope you understand, but if I hadn't he would have loathed me.'

And before going to bed that evening, Era was sent to say good night. It was purely formal but we faced each other as equal to equal, without affection but without antipathy; and the next day I got him interested in a story about *ikan hi* (sharks) with *gighi besar* (big teeth), and we soon became real friends.

Anyone who is against whacking children under any circumstances, because it is brutal, is possibly on the wrong track. Brutality is wrong, but a whacking that stings sufficiently to impress its lesson need not be brutal and may be actively constructive. Sometimes even angelic children can be the most objectionable, beastly little wretches, deliberately and cold-bloodedly bloody. In the first place I never see why a grown-up should submit to that, simply because having the power to stop it, it is considered unfair to use it. In the second place, if the child gets away with it he will quite justifiably feel contempt, because he knows that the grown-up has preventive power but is either too stupid or morally weak to use it. In the third place, such grown-ups luxuriating in their 'morality' do the child great harm by deliberately causing him unhappiness—to allow the child to engender contempt automatically precludes him from experiencing either love or happiness.

This is exactly the same in principle and only different in degree on a national scale. The recent case of the Egyptian Nasser tearing up an international treaty and defying England and France, believing that the fear of Russian intervention would preclude retaliation, is identical to that of the child spitting on me and believing that my fear of his fond mother would preclude me from my own self-defence.

In the national case many of the world (including many Englishmen, to their shame) cried, 'Aggression' as if aggression is always wrong. This quite overlooked the fact that tearing up an international treaty is just as much an act of aggression as the landing of troops on foreign soil; it is aggression in another form, just as beauty in a human face or beauty in literature is expressed in different forms but is still beauty.

'But,' say the moralists, 'one act of aggression does not justify another, because two wrongs do not make a right.' That is not always true, because one wrong committed without repentance and unchecked is going to lead to many more, which will total far more harm than merely one other to neutralise the original.

The political situation in Indonesia was interesting, partly in comparison with other countries which had recently achieved independence, and partly because despite being one of the most richly endowed areas in the world it was a financial wreck. No Indonesian I met would discuss Russia, but several freely discussed the alternative solution of help by the USA. They expressed the view I had heard in India and Ceylon also, but first let me stress that the USA loses much of its potential power in the East, as a champion of democracy, by the manner in which she handles her own colour problem. And so the world loses also.

The USA is seen as a country with a very high standard of living, tremendous productive capacity, a saturated home market, and surplus capital for investment—which can only be invested abroad if that high standard of living is to be maintained. And no government which does not maintain it will stay in power.

The Indonesians I met therefore felt that the acceptance of USA capital could lead too easily to economic control, because of the American incentive to exercise it; and economic control so easily becomes political control, meaning the loss of their independence so recently and so hardly won. The cultural aspect was also a factor, as it was feared that the American way of life might swamp local culture, on which their whole philosophy of living is built. Coming from an undeveloped country myself I could appreciate this outlook only too well.

During these months, apart from writing, I kept *Sheila* in commission and worked on board, re-caulking the decks, re-rigging, and completing various other tasks. Fien took me to a firm of Dutch sailmakers, saying he wanted me to see their work. The manager showed us around and the work was excellent.

'Then you'd better order a new mainsail,' said Fien. 'You said that your present one had given much trouble.' The manager had already told Fien that a sail for *Sheila*'s size would come to the equivalent of about £75.

I declined, of course, until Fien said that several of them had already clubbed together and had decided to get me a new mainsail. This is only one instance of help received. Three young Dutchmen who planned to sail a converted life-boat to Australia the following year also helped me with ship's stores, gear, and technical work.

Although I spent much of the time thus busy with *Sheila* and with my writing, the greater part of it was spent socially in accordance with the custom throughout the East. Club life. Most of the Dutch were making large sums of money which was largely worthless even if they could have got it out of the country, and so entertainment was lavish and almost continuous. Sometimes I hated it; it brought that clamour, 'I'm having a wonderful time—but?' What is it that gives that feeling of lack, of dissatisfaction, of need? And so I knew it was time for me to move on, and decided to go south to Australia.

28

Bali

For many years it had been one of my ambitions to go one day to Bali, and so it came about in the eventual way that cherished ambitions often do.

I was prepared to be initially disappointed, because so often whenever a place on this earth is found to have such beauty as will attract large crowds of tourists, it is only a matter of time before the large crowds of tourists destroy it. The simple unspoilt Balinese, anyway in the vicinity of the large modern Bali Hotel in the capital Den Pasar, had become shrewd and persistent commercials; there were also beggars and cadgers, pimps, perverts, and prostitutes, where before these had not existed because there was no demand for them. What can excuse the human race for creating evil within itself in this manner?

For the first few days in Bali I rushed about in a car seeing all that it was said ought to be seen, and slowly became aware of a growing sense of the disappointment I had expected. So in the second week I went to a beach and sat still, lay in the sun and the surf, and only went back to the hotel in the evening.

In this way I missed much that the others saw, but I was seeking something different; it was like returning to the Louvre in Paris. You can hurry through and see all that is to be seen and thereafter say, 'I have seen the world's masterpieces—' So what? You will not even know why they are masterpieces. In the same way in Bali, I sought not their carving, painting, music, nor even their dancing, but rather that which lies behind these and gives them birth.

Much of my time on the beach I spent writing. And outside my writing there was much to see—the slim outriggers rose and fell on the far edge of the breakers covering the reef; at high tide people dug in the sand at the high-water line, following crabs to the bottom of their tunnels, to string them on a thin sliver of bamboo and take them home to eat; at low tide the reef was dotted with figures fishing in the rock pools, or collecting lumps of coral used in building their homes; in the evening the fishing boats returned home through the surf, and long strings of girls strode beautifully over the sand carrying tall baskets on their heads. Once I slept on the beach all night and watched the sun awaken the island; and having found the wonder of the dawn at sea, I found it now more deeply where it awoke so many others with myself. It was like the remembered chord in a forgotten piece of music, which revived to spread its beauty outwards to embrace the whole piece.

In this mood it was safe to return to the tourist shops, to take up a carving and feel the texture of the wood beneath my fingertips, to see whether the cuts were hurried or fine, to study the expression on the face and try to feel what, if anything it conveyed; to put the figure down and watch it, seeking to find if the artist lost in his work had breathed life into it, or whether the artist, thinking of a wealthy tourist, had not.

I saw several dances organised and staged at the Bali Hotel, and these were beautiful but again were not what I sought. I felt it better to go to a village to see dancing for the people's own amusement, where the mass appreciation of it creates the atmosphere from which it is given back, where I could find that unity in which giving and receiving become one.

This eventually happened quite by chance. I had spent all day on the beach, and had stayed late to watch the night come in from the western sea and cover the island. It comes so suddenly in the tropics. On my way home in a borrowed car I heard the tinkling music and the muffled drums of a gamelin orchestra, so I left the car and made my way on foot down a dark lane towards the music.

The village houses were in darkness, but an open space of beaten earth was lighted by bright oil-lamps. At one end was a wide curtain hung between two poles, and at the other were the musicians. Some were squatted behind their gamelins, others with long drums across their knees, some with tiny flutes, and some with lengths of hollow bamboo which they beat endwise on the ground. On either side and behind the musicians were rows of faces flowing back into the darkness, until only an occasional movement was visible. I passed into the crowd and squatted down under a tree behind the orchestra.

The gamelins took on a more insistent note, still soft and rhythmic but more compelling, directing my eyes to the curtain at the far end of the dancing floor. Soon these parted and a young girl stood in a typical pose which I recognised from Hindu dancing, expectant, waiting. She was dressed in a beautifully coloured sarong, her waist and breasts tightly bound in gold cloth, and a crown of flowers built into her dark hair. In one slim hand she held a fan, and it seemed that she was waiting for the music to draw her away from the curtain. Then she advanced into the open space, her feet moving to the music, which flowed restrained in her body until released outwards along her arms to the slender hands and delicate fingers.

I was unaware of the music and yet absorbed by it as it flowed through the girl's movements. Twice she moved into the crowd and touched a man lightly with her fan—her invitation to dance (this is an honour, an insult to refuse). The man chosen went to the orchestra where a short sarong hung, twisted it about his waist, and danced with her, never touching as in the West, but advancing, retreating, and turning with her as in a ballet.

Later she returned to the curtains and waited until the music allowed her to depart. There was no clapping, and during the interval before the next dance the music continued, more subdued perhaps.

Again the insistent note grew into the music, and soon another girl stood poised between the curtains. After dancing for some minutes she moved towards the crowd beside the orchestra; the squatting audience shuffled aside a few inches to let her pass among them. As if unaware that they existed she never looked down. When she was some way into the crowd, with deliberation but without haste she turned and soon stood before me, still moving to the music, her dark eyes strangely promising, strangely dispassionate, looking directly into mine; from one of the flowing movements of her arm her fan descended and for a fleeting moment rested on my knee. She continued swaying before me, until without haste, without glancing down to break the spell, she moved back to the open space, waiting.

I rose and moved through the crowd, past her to the orchestra, took up the sarong and wound it around my waist as I had seen others do. Moving to the music I advanced towards her, part of my mind terrified at the spectacle I presented, part listening to the drums, but all of me very aware of the girl before me. She moved away as I advanced, but soon she smiled faintly and swayed towards me until our bodies were almost touching; her eyes never left mine. She seemed to feel the response in me and moved away again, still faintly smiling. It was the same chasing game they played in Siam.

Later, safe under my tree once more, I watched her closely. She had beautiful features but her real beauty came from something inside, part of her, yet beyond her. There was a child-like candour and innocence, but a maturity too, not such as can be achieved in a lifetime. Behind this was a very definite awareness of her femininity with all its power and delight, but she seemed detached from it. While yet using it and even surrendering fully to it, she was never enslaved by it.

I wondered whether an artist could not capture this something from within, and whether it is not this that is the goal of all art. Is it this that makes a masterpiece stand out through centuries of changing fashions and trends? I knew that it pervades everything, and that, once seen, the outward form through which it is perceived becomes unimportant; seen in a painting for instance, the actual picture itself then becomes nothing, being merely a means to an end.

There was the portrait of a man's face elsewhere in Bali, on seeing which I had shuddered and said, 'God, how terrible,' but had been drawn back to it again and again until I understood.

It was the face of a young man old in vice, ravaged by it, enslaved by it, and only in the eyes was shown the agony of the struggle against it. There can be no struggle against evil unless there is awareness of equal good, because without opposition no struggle can exist; and so this painting portraying one portrayed both, which cancelled out to leave the reality underlying all things.

The night's show ended well after midnight, and the crowd dispersed into the darkness beyond the range of the two bright lights. With others I found my way down the narrow bamboo-enclosed path to the road, strangely disturbed by all I had seen and the thoughts induced, knowing I would not sleep.

The figure of a girl stood beside the locked door of the car. She wore an ordinary village sarong, a cheap cotton blouse loosely covered her shoulders, and my night-accustomed eyes saw the girl I had danced with. It seemed a long time I looked at her, unwilling to speak; and then I asked her if she would like to come to the beach.

I drove the car off the road and through the tall palms to the very edge of the sand, where we stepped out of our clothes and ran down to the sea. The water was full of glowing phosphorescence, refreshing and cleansing after the hours of squatting tight-packed in the crowd. The warm night air dried us.

She stirred sometime before dawn, kneeling beside me, sitting back on her feet and stretching; from where I opened my eyes her slender arms seemed to reach high into the tops of the palms. Then she leant forward, a hand either side of me, bending down to find my eyes open, and her long dark hair fell over me in a cloud to hold the night a little longer.

When dawn brought light to the curving beach we ran down to the sea and splashed in the surf, dived to the sandy bottom in the green-lit water, and her long hair floated about her like a mermaid's; and when the first sun touched the tops of the palms we left the sea—and I remembered another morning beside water, when the same sun had touched the highest buildings and flowed slowly downwards to the bridges of the Seine.

But now we hopped about on the sand with short stiff-legged jumps to shake the drops of water from our bodies. I laughed because I felt wonderful; she laughed, doubled up and helpless with laughter, because I looked ridiculous—from my insteps to the roots of my hair my body was as brown as hers, except for a vivid-white triangle fore and aft where my bathing Vs worn at sea had ever denied the sun.

That was a beautiful morning, and we both felt a happiness which sprang from a deep gratitude, from a harmony we found in each other, for the sunshine on the coast-line, for the cool tinge in the warm air playing over us, for the taste of salt on our lips, for our laughter; not consciously of course, but these things were there and we were grateful, she to her God, I to mine; yet these things came from the same God.

Since that time I have often wondered whether the Garden of Eden has in fact no place in Time or geography, and whether the story of Adam and Eve is not written every day. They knew beauty in all things around them and knew no lust in their nakedness, until the Serpent brought desire and the Garden vanished. And if we wish to regain what Adam threw away, we must learn again to love beyond desire.

And so I left Bali, but I shall never forget.

In early January 1954 I said good-bye to friends in Surabaya. Kitty, Fien and Era came out in the club dinghy to put me on board *Sheila*, and I could not grasp Time nor any feeling beyond a numbness, until *Sheila's* bow was turned to the harbour entrance and the club was behind me. It was the saddest parting in all my voyage.

29

*The south-east trades—Sheila springs a leak—emergency rationing—
a dirty bottom—communism—a hard decision*

The 1600-mile voyage to Fremantle I regarded in two phases; the first from Surabaya via Bali Strait to North-west Cape Australia, and the second from there down the coast to Fremantle. Two other yachts in the past had attempted this voyage and both had turned back, one leaking so badly she only just reached Bali and was driven ashore. The reason for defeat in both cases was the south-east trades, which make the voyage from Java to Australia one hard plug into consistent head winds and seas, a severe strain on ship, sails and crew.

The failure of the two other yachts meant little to me, because failure is always inconclusive—otherwise, after all the previous failures to climb Mt. Everest why did Hillary and Tensing even attempt it? I estimated that *Sheila* could do this voyage in eight weeks, and I laid in the usual forty gallons of fresh water and six weeks' supplies of food. It would have been wiser to have laid in eight weeks' supplies and a fifty per cent reserve, but I cut it down because, like everything else during my eight month's stay in Surabaya, the supplies were a gift from Kitty and Fien. They had begrudged me nothing, and would gladly have doubled my supplies, but I assured them that six weeks' was ample. This may be a psychological kink in me, but it made me feel that this was the only way in which I could fulfil my gratitude to them although they might never know of it.

The first few days and nights along Madura Strait brought the usual doldrum conditions until I entered the narrow reef-fringed passage separating Java and Bali. By dusk on the seventh day out from Surabaya only the high peaks of Bali were visible astern, and when darkness fell I said good-bye to Asia. At dawn I was alone with the sea, headed south for Australia.

I picked up the south-east trades about 100 miles south of Bali, where these winds blow night and day under clear skies for months on end. *Sheila* sailed herself with no attention from me, ploughing her way south at about ninety miles a day. During daylight I read, wrote, worked out navigational problems, or just lay in the sun on the foredeck. And through the hours of darkness I left *Sheila* under full sail to carry my sleeping body safely across the miles of empty ocean.

A week after I left Bali the weather suddenly blew up into a short sharp gale, and demanded a change to storm canvas. It was that night I first awoke to the loud sloshing of water over the cabin floor. I pumped it out and went back to sleep, only to be similarly awakened an hour later. *Sheila* had sprung a leak, badly.

When the weather moderated the stream was easily traced into the fo'c'sle, and after dismantling and lifting out the lavatory I uncovered the leak either side of a rib. As *Sheila* plugged into each sea and the stays converted the forward tendency of her mast into a downward thrust, two solid jets of water hissed in through her planks. Perhaps I had set up the stays too tightly, but the real cause I only learnt months later; the use of brass screws for an under-water repair job in England.

Little can be done to remedy a leak from the inside, as the pressure from the outside rejects any packing; and to use force is unwise, because if the leak has been caused by the teredo worm such force may drive a hole right through the planking.

No marine glue would hold to the wet wood, and so I finally made red-lead sandwiches between dry pieces of cloth, packed these over and around the leak, held them in place by building a box over all, and screwed it down tightly. This achieved little, and was probably not even an insurance against a sudden burst inwards (which was what I really feared), but it was good for morale because it was all I could do.

At this time I seriously considered turning back, once again struggling to draw the line between wise assessment and stubborn persistence, between courage and stupidity, faith and logic, submission and defeat. I was very afraid that the leak might get worse and I could have turned with the winds, eased the strain on the planking, and been safe on land within a few days; Kitty and Fien would have received me again. But I could not know definitely that the leak would get worse, and how many failures have been due merely to a pessimistic assessment of what might happen? If I evaded the payment how could I expect the reward? So I continued south.

Thereafter *Sheila* needed pumping out every hour, at shorter intervals in rough weather, longer in calm; and that went on for the rest of the voyage, night and day, storm and calm, under conditions of which I fortunately had no inkling then.

And so *Sheila* continued to punch steadily into the south-east trades, headed for Australia. Her new mainsail was excellent and she seldom needed my hand on the helm, which is one advantage to the single-hander of sailing to windward. Being off the shipping lanes I slept at night, except for waking every hour to pump out. If ever I slept for more than an hour, the sloshing of the water against the side of my bunk awoke me, and sometimes took as many as 2000 turns of the pump handle to remove. I always counted the turns so that any increase would give some warning if the leak became worse.

This area of the chart half-way to Australia showed depths of over 3000 fathoms, and at night twin glowing orbs of phosphorescence made me wonder whether these might be the eyes of giant deep-sea squid. I had read that these monsters come to the surface at night in search of food, and tales had been told of them climbing on board sailing vessels becalmed and plucking men from the deck with their huge tentacles. *Sheila*'s freeboard seemed to be very inadequate during the long dark hours at the helm, and I kept the axe beside me.

Three weeks out my radio succumbed to the dampness, depriving me of accurate Greenwich time which is essential to find longitude. The daily rate I had worked out for my wristwatch was not sufficiently reliable over long periods without checking against radio time signals. I was not terribly worried, because I thought that my watch would be sufficiently accurate to get me to sight Australia, whereafter I intended to travel down the coast to Fremantle and then be able to do without longitude; but it was this lack that later so nearly brought the voyage to disaster.

The same day that the radio died the fresh-water pump failed from the after tank (which I had used for the first week out); after being cleared it brought up nothing but filth and mud.

This was a new tank which I had cleaned, installed, and filled myself, and I could only assume that the filth had been inserted from the filler on deck the night before I sailed (while I was sleeping ashore), and had taken the three weeks to work down to the outlet to the pump. Perhaps some embittered locals had taken this form of revenge against the hated whites, and far out at sea I realised that this was neither more stupid, nor less criminal, than burning the many sacks of rice that I myself had on occasions burned in Malaya when patrolling squatter areas. Anyway, because of this sudden loss of fifteen gallons of drinking water I immediately put myself on a ration of two pints a day for all purposes.

At this time also, and for a reason I shall never understand, the kerosene fuel for the cooking stove and lamps began to give trouble. It was stored in the usual container and up to this date had given no indication of being affected, but perhaps it had also been polluted; now it would barely burn in a hurricane lamp and formed a hard deposit in the stove jet. I managed to keep this clear for several more weeks by a careful use of the spare prickers and two spare burners, but eventually that fuel was to render the stove useless.

My wristwatch must have been keeping its normal rate, because North-west Cape appeared out of the sand-haze on the following morning exactly as expected from navigation. It is about half-way to Fremantle and had taken twenty-seven days to reach, say four weeks.

Immediately on leaving Surabaya I had put myself on a ration, eating about three-quarters of what I did normally. This soon became a habit and was then almost unnoticed, but at North-west Cape I decreased this to half-rations, which was not so severe as to bring extreme pangs of hunger but harsh enough to make me continually aware of hunger.

And the reason why I enforced this ration was because I noticed that *Sheila*'s bottom was becoming very dirty.

It is not generally known what a tremendous effect a dirty bottom makes to the speed and manoeuvrability of a sailing ship, particularly to windward when a dirty ship will only sail at right angles to a head wind, rather than the normal 45 degrees into it. The marine growth concerned in this case was the goose barnacle, a small shell-fish shaped something like the New Zealand pipi, about an inch long, and growing on a rubber-like stalk also about an inch long. A few weeks later I doubt if there were two square feet of *Sheila*'s under-water hull not covered by this pest.

It is hard to understand how this growth could have taken hold on a ship under sail, particularly moving from tropical waters into cold, for such a change of water-temperature will kill many marine growths. *Sheila* had been freshly painted just before leaving Surabaya, and the paint had been a well-known international brand kindly supplied by the manager of a Dutch shipping firm. He had himself issued a new tin from his store, and although I had been asked not to scrub *Sheila* down and paint her myself, for fear of upsetting local labour, I had stood by while these two tasks were done to my satisfaction.

Of all the factors which by then had accumulated against me, this was by far the most serious. I could deal with the continual pumping out, the short rations of food and water, for a reasonable length of time; but a dirty ship would tend to prolong that length of time beyond my endurance (and very nearly did). Again, a dirty bottom demanded hard sailing to give *Sheila* the power to move at all, but the leak demanded careful sailing to avoid any increased strain. Here again all logic and seamanship told me to go back, to take the strain off the ship by turning with the wind and make a swift easy passage back to Bali, but to hell with that.

Short sharp blows of gale force now became more frequent. The rough weather meant more work on the pump, not only because *Sheila* leaked more with the greater strains imposed on her, but also because of breakers which sometimes covered the ship, filling the cockpit and spurting through the closed skylight. To ease the strain I lay hove to, which meant *Sheila* drifted to the north and lost precious miles so hardly gained; it also left me nothing to do but sit in the cabin and pump out every half-hour, so between whiles I read and one book was by a leading British communist.

The idea of communism appals me, as does any other imposition which subjugates the individual to a human assessment of what is good for him, whether he likes it or not. I had been in full agreement with John Stuart Mill, who thought that society should be organised with the object of affording the individual the greatest possible liberty. Communism evidently thinks the opposite, that the individual may be imposed upon without limit for the good of the State—whatever that is. So I read this book to try and understand the mind of one to whom that made sense.

The communist worship of materialism did not provide so much interest, because here there is little conflict. It seems to me that those who place a primary and supreme importance on the attainment of a higher standard of living worship materialism just as fervently as the communists.

Nor did there seem to be such conflict in the non-material arguments. As a smashing argument to support materialism and banish religion, the writer said that modern science had proved that Christ achieved His miracles of healing not by divine power but by psychology. Well for heaven's sake what's the difference? Psychology is only understanding and that surely is divine?

Sunday was a lovely day, bringing peace after storm as it so often seemed to do, and at dawn I found a fat flying fish stranded in the cockpit to solve the problem of breakfast. I spent most of the day cleaning up below, and brought clothing and bedding on deck to dry. Then I poured some detergent and disinfectant into a bucket of sea-water and scrubbed out the galley complete, until it smelt like a hospital and must have been nearly as clean.

The following evening again brought a rising wind, and in trying to give *Sheila* the power she needed I held on to sail too long; when I finally reefed there was a long r-r-r-r-rip, and the main tore for nine feet along the leech. We lay hove to for the next three days under storm canvas, and when the weather eased I bent on the spare sail, the old one which had sailed me from Aden to Surabaya, but that very night it blew to tatters in a squall just like you read about in books. It was not worth mending, being rotten, and so I cast it into the sea.

It is difficult mending a big sail in a small cabin, particularly when all has to be laid aside every hour to pump out. It took me three days to stitch the two long edges together evenly, and to sew a patch over all.

Even after three weeks the pumping chore was beginning to tell. Combined with the short rations of food and water and the strain of bad weather, it brought on a state of nervous depression and fits of temper which I have come to know are a sure sign of over-fatigue. Children also develop the same symptoms when not sent to bed early enough! My diary records:

> ...afraid of over-sleeping, and awaking to find so much water in *Sheila* it nearly kills me to shift it. It's hell going to sleep very tired, and knowing that in just one hour I will awake to that fearful sound of water sloshing against my bunk—what wouldn't I give for 8 hours undisturbed? I could never understand before why the Nazis used the denial of sleep as a form of torture. Sometimes in those seconds between sleeping and waking I panic in the utter blackness of the cabin, thinking I'm shut in a box which is sinking, not knowing nor seeing any way of escape.

By mid-February I was 200 miles south of North-west Cape, an average gain to the south of only ten miles a day since sighting Australia. The same weather could be expected for all of the 500 miles remaining to Fremantle, meaning another seven weeks at sea. The idea was appalling.

I was then off Bernier Island, outside the huge indent on the western bulge of Australia known as Shark Bay, inside which lies the port of Carnarvon. The sensible thing obviously was to go in there to clean off *Sheila*'s hull, repair the leak, and replenish food and water. So with this in mind I sought all available information in the Pilot Book and a large-scale chart bought for just this possibility.

The chart showed that to reach a depth of eighteen feet of water the port's wharf was a mile long, meaning that with such a shallow shelf there would be no slip-way for boats, and the soft mud bottom put beaching out of the question. Nor did the Pilot mention any repair facilities, and where these existed they always were mentioned. But even if they did exist I had no money for slipping and shipwright's work. Local jobs would be available by which to earn money, but during that time *Sheila* would have to lie at an open anchorage at which I knew many pearling luggers had been lost in hurricanes. It was then the hurricane season (the Australians call them 'Willy-willies'), and *Sheila* was safer at sea than at such an anchorage.

As repairs or slipping were not therefore feasible in Carnarvon was it worthwhile going in for food and water only? My decision was that no, it was not. Already my nerves were affected to some extent, and I knew that having gained the security of land I would never have the will to leave it with *Sheila* in that condition. The security in port thus became a greater menace to my reaching Fremantle than the hazards at sea.

Apart from all that I believed I would get to Fremantle, and this again brought me face to face with an old problem—on one side safety and logic, and on the other probable disaster and belief. But just what is belief? I had said many times in all sincerity that I believed this or that, but that belief became mere lip-service in a situation like this, where life and death hung in the balance.

I only knew that I had a quiet persistent feeling that I would get to Fremantle, and it is stupid to have a belief if you are afraid to act on it; in fact if you are afraid it means that it is not belief otherwise you would act on it. My thoughts were confused but anyway with a bad leak, dwindling food supplies, very little water, rotten sails, and a barely manoeuvrable ship, I headed 400 miles out to sea, away from land and shipping.

30

*Belief—help from Kwan Yin—the condensing plant—
the problem of fire—ships passed—the last sunset—underwater landing*

This decision was not taken without a careful assessment of what it involved—not to judge whether the belief held chances of practical possibility, but to plan how best to harmonise with that belief. Just because I believed I would get to Fremantle did not allow me to sit idly on my bottom while some power sailed the ship; in all probability it meant hell and high water.

Later, much later, this suddenly gave me the understanding of belief I had sought for so long. Belief, it seems to me, is not of one thing or of another, but is something in itself complete, which we may manifest and so receive its power. This is not done merely by saying we believe this or that, or by wanting this or that, but only in the belief that it will be so have we the strength to make the sacrifices and endure the ordeals that will make it so.

With *Sheila*'s under-water hull as it was it was no use trying to battle against the persistent southerlies which blew hard up the coast, and in searching the wind chart for an alternative I had found that 400 miles out to the west away from land there were some westerly and even nor'-west winds. It would be feasible to work out to these across the southerlies, and then use the westerlies to blow me south to the latitude of Fremantle. According to the Pilot Book the northern fringe of the Roaring Forties would then pick me up and drive me into my destination. This plan meant going 1300 miles to cover what lay only 500 direct, and if the expected winds did not blow I would be far out in an empty ocean miles from any help, and where no one would think of looking for me; but it was a risk which had to be taken.

Three evenings later the bowsprit broke when a steep sea smashed into the belly of the jib. The tangled mass of torn sail and splintered wood flogged away to leeward, releasing the topmast fore-stay (a main support for the mast). I worked desperately to rig the spare fore-stay on which the storm jib sets to the stem-head. This only took a moment and saved the mast, which was flopping about like a drunken sapling, from crashing overboard. I recovered the wreckage of the bowsprit and lashed it on the foredeck, because it was then too dark to do anything about it.

Thereafter *Sheila* became a stem-head rig until I could make a new bowsprit; as it was not a rig for which she was designed it decreased her speed appreciably. I immediately reduced the water ration to one pint a day, which on the fringe of the tropics was no pleasure.

That same day I inspected and made a list of all the food that was left, and decided on a ration which would draw this out to last thirty days, adding a note for my own benefit, '…but pray God it doesn't have to.' It was the barest daily minimum on which I could hope to survive.

It was not long after this that the meagre rations of food and water, the never-ending pumping out and the lack of sleep began to affect me. And my diary records the first reference: 'Passed out for a few minutes—lack of food and water.'

These fainting fits were to become more frequent, and although harmless in themselves (I could always feel them coming and if on deck get into the safety of the cockpit), they might continue straight into a long deep sleep lasting for hours. In this time *Sheila* would have filled and sunk, or have taken in more water than I had the strength to remove.

In bad weather the water reached the floor boards in one hour and then took twenty minutes to remove, leaving me exhausted; she was taking in water at about one-fifth of the rate at which I could pump it out. And I had run out of cigarettes when I desperately needed their comfort, because at that time life held no other.

It was often an unreal feeling being in such distress so far from any help, particularly at night sitting on the edge of my bunk in the pitch darkness after pumping out. Owing to the kerosene failure I had not had a light for weeks. The little 'juice' left in the batteries I saved for the Aldis lamp in case it was needed for signalling.

I searched the ship twice from stem to stern for cigarettes, looking in every cranny and locker, hoping that perhaps some had been misplaced. This was unlikely because I always stored them in one particular locker, just as all other supplies had their one particular place.

A few days later something drove me to search the ship a third time, knowing it was quite useless and wondering whether I was not going mad. In the laundry locker was a small bundle wrapped in a towel. It was a jade statuette, said to be 800 years old, of Kwan Yin the Chinese Goddess of Mercy, and it had been given to me by someone who loved me. And behind the place where the wrapping had lain was a whole carton of 200 cigarettes. I am quite sure that I had never put them there, no one else had been on board since I had lain the statuette there for safety, and why had they not been exposed to view on previous searches? There must be an explanation, of course, but perhaps it was only my need.

After three weeks' sailing to the west I found the fair winds to take me south. The weather too improved, and running before the wind *Sheila* leaked less and allowed longer hours at the helm. I drove myself to the limit to use every fair breath, as my diary records:

> 8th March. Sailed all day, spinnaker up from noon and kept it there all through the night.

> 9th March. Slept from 6 a.m. to 8, upped spinnaker till noon. Took noon sight and suddenly felt very dizzy—and then went crazy and drank a whole pint of water.

I had reduced the daily water ration still further to one small tea-cup, about one third of a pint, which I took at 6 p.m. each evening and not one second before. After looking forward to that drink for twenty-four hours, ladling it out of the can of eight or nine pints, enough to quench my dreadful thirst, but holding myself to just that one small cupful, brought the harshest daily test I have ever endured.

It was too great a risk to let myself remain so weak, and with only two and a half pints of water left on board I knew I had to lose time at the helm and experiment with condensing water, in which apparently lay my only hope of survival.

Condensation depends on converting water—sea water in this case—into steam in a boiler, and leading it into a container cool enough to convert the steam, purified of the salt content, back into water.

During that first day tests looked hopeful so I continued experiments far into the night, at the end of which time I had produced one-third of a pint. Next day, after further adjustments I made a whole pint, which I promptly drank. For the rest of that week I lived on precious drops while experimenting with other tins and systems until near-perfection was attained.

An old pressure cooker sealed with marine glue became the boiler. A short length of rubber tubing made a flexible joint (to allow for the movement of the boiler on gimbals so I could distil however rough the weather) to a copper pipe taken from the water cooling system of the engine, which led the steam into an empty two-pound margarine tin weighted down in a bucket of cold sea-water. This water was changed by deluges from the bilge pump as soon as it got warm, which it did surprisingly quickly from the heated steam entering the tin.

The fair winds died and during the calm which followed I made a new bowsprit out of the spinnaker boom. I believed that this would help *Sheila* more than the spinnaker, as the winds from astern were comparatively few; and not only was I becoming too weak to handle the big sail, but once it was set I had to stay at the helm for long periods. *Sheila* leaked less before the wind, but after long stretches at the helm the water was half-way up the bunks and was then very exhausting to remove.

The new sprit was made exactly the same as the old one, but not being such a thick spar I ran rope 'whiskers' aft over spreaders and kept them taut with blocks and tackles. It did good service for the rest of the trip.

Sheila was now so dirty below that she was barely manoeuvrable, would not go about at all, and only just wear. I had heard that some of the greatest voyages of discovery in the Pacific were made because the old galleons running before the Trades got so dirty that they could not turn about or wear, and just had to keep going. I can well believe it.

It was no use trying to remove the barnacles by keel-hauling a four-gallon tin—the barnacles merely swayed on their rubber stalks and the tin slid over them. I did manage to clean as far down as the turn of the bilge with the dinghy oar, but this made little difference to *Sheila*'s performance.

On Sunday, 21 March I wrote:

Ten weeks out today—I hope people aren't worrying too much, nor trying to incite the Australian Navy and Air Force to look for me. They'd probably search up and down the coast, whereas I'm now about 500 miles offshore.

10.30. No wind, nearly crazy with hunger, tried to sleep to by-pass time and feeling. Dozed for two hours, pumped out, dorado alongside too tempting—I hate killing these things, partly because they are so lovely and companionable, partly because I hate killing anyway. However I gaffed one, and as it died in the cockpit it sicked up two big flying fish. It couldn't have had them long as the digestive juices had only removed the skin and fins—fried them both, and they were beautiful, really beautiful.

I lived on dorado whenever catchable, and each fish kept me for two days. The two large fillets on either side fried, as did the spine with whatever flesh remained after filleting. Boiled, the head rendered a surprising amount of delicious meat and left a nourishing soup. I also raided the insides for strange looking organs which proved to be edible, and these with any tiddlers and squid found in the stomach also went into the soup. Any big fish inside, such as flying fish, even if half-digested, were fried.

I never caught a dorado while there was still any of the previous catch on board, and sometimes with only one meal left it was tempting to gaff one while I could before that last meal was finished. The fish often disappeared for three or four days at a time, or I was unable to hook them in rough weather. But I never did hook a second before finishing the first, and nor did I waste any portion of any fish killed, believing that only thus was I justified in taking their life to support my own.

Sometimes this lack of 'foresight' led to periods of great hunger, and one night after pumping out I sat on my bunk, pressing my hands into my stomach against the pain and said aloud, 'God, I'm hungry.'

In that very instant there was a loud smack on deck and a frantic clattering, and I hurried above to grab a large flying fish which had hit the cabin-top in flight and was trying to flip back over the side. Strange things like this happen at sea; they happen on land too, but there they are called coincidence.

The stove finally succumbed to the bad fuel and I could keep the last jet open no longer. Normally I could do without fire but now I had to have it for condensing water, and so I racked my brains for an alternative method of producing it. I did not think a small wood stove of the size possible would be sufficient.

With a firewood shortage in the desert the troops had used a few inches of sand in the bottom of a tin, poured a little petrol on top, and so produced a hot flame for a surprising length of time. I had about fifteen gallons of petrol on board for an engine which was valueless in that expanse of ocean, and eight pounds of rice which was useless as food without the fresh water in which to cook it. This rice acted as a substitute for sand, and although while burning it formed a crust on top which had to be removed from time to time to save the flame, it did work surprisingly well. A large jam tin slung in the gimbals made a suitable stove.

This system had two big disadvantages. The first was the danger of fire—there were two bad explosions in both of which I lost my eyebrows—and the second was smoke. Unfortunately I had previously mixed oil with the petrol ready for use in the two-stroke engine, and so the volumes of thick oily smoke soon coated everything in the galley and saloon in a greasy blackness.

Another Sunday came and I celebrated the success of the condensing plant by bringing out my secret reserve of water—a bottle of distilled water originally brought on board for the batteries. With the tiny bit of coffee left I made a full pint mug, which I drank lying back in my bunk like any civilised human on a fine Sunday morning. The rest of the day I spent at the helm and there was little need to leave it for food, because there was almost no food left—about three pounds of rice and half a bottle of soya bean sauce.

Towards the end of March I had covered the 500 miles south to the latitude of Fremantle, and with a thrill of joy I pushed the helm over and turned east, headed once more towards land.

The condensing plant was by now producing two quarts of water for three hours' work, and so considering the food shortage and having enough water in which to cook rice, this became too valuable to use as fuel. I made a chip stove out of a two-pound margarine tin, slung it in gimbals, and it worked far better than expected. It used a lot of wood, and as it was very sensitive to damp wood, *Sheila*'s beautiful mahogany locker doors and saloon panelling were ripped out and chopped up to feed the flames.

For the first time in my yachting history I spent the afternoons chopping firewood. The larger pieces I cut up with the axe and then split into kindling size with a knife, and it was only a matter of time before this slipped, cutting the top of my thumb to the bone, an inch down one side and over the top to the far corner of the nail. It gaped open, spurting blood, and obviously needed stitching, but with only one hand left, this was happily out of the question.

I cleaned that cut very carefully, digging ruthlessly inside after tiny chips of wood and specks of dirt, and thereafter each morning and evening I washed it in disinfectant and bound it up with a generous wrapping of cotton-wool to guard it against knocks. It eventually healed successfully many weeks later, and gave no trouble beyond creating one more daily chore, and being inconvenient when changing sails and gutting fish.

In those southern latitudes (32 degrees south) flying fish were scarcer and so were the dorado which fed on them, and so I who fed on the dorado became hungrier. There were days when I had to go without any food at all, and the unpleasant ache drove me to search my brains for ideas on how to raise my standard of living. This search seemed a waste of time after the hours already spent thus, but I have often found that in persisting with some problem already deemed insoluble, an incredibly simple answer suddenly pops into mind. And so it did here—the barnacles!

I lay along the deck, gripped the gunwale with one hand, and lowered my body over the side until with the other I was able to gather the clusters of shellfish growing below the waterline. I had no idea if they were edible, but there was only one way to find out. Boiled with a dash of soya bean sauce they were delicious and also proved to be very sustaining. The rubbery stalk was no good and was snipped off, like snipping beans, but having drunk the soup I then cracked open the shells and sucked out what was left. This diet kept me on several occasions when the dorado failed to co-operate.

Later, when the barnacles within easy reach had been plucked, this marine gardening had its hazards. I never went over the side after them because I was afraid of not being strong enough to get back on board, and also of sharks; instead I tied a rope round my waist, and with another to hang on to below the gunwale, I lowered myself head and shoulders under water to reach well below the turn of the bilge, but with my legs still on deck.

One still day as *Sheila* lay becalmed I came up for air to hear a loud sustained noise of broken water. It was an unnatural sound in that stillness, so I hastily dragged myself on board. Then I saw the shark, a huge brute, coming towards *Sheila* in a wide, fast curve, his big dorsal fin bending as it braked against the turn, and it was this that made the noise. If I had been fit and able to hold my breath as long as usual, that shark would have got me with my head still under water.

The winds were mainly good, from the south-west, and a week after turning east my DR put me very close to Rottnest Island, which is just outside Fremantle. At this time I got my latitude each day at noon, but not having GMT did not know my longitude; so I knew I was level with Fremantle and heading towards it but I could only guess how far off I was.

That night a ship passed, which was some confirmation of my DR, and I let her pass without signalling to report my position or to check on it. This dreadful neglect may have been due to the inefficiency of fatigue, because I fainted at the helm soon after she had passed. Or it may have been that, sure of my DR, I thought I was so close it was not worth bothering anyone else; or it may have been because I never ask for help unless it is absolutely necessary. A professional seaman had once said to me: 'You ruddy amateurs are a ruddy nuisance. You push off to sea in your little boats, lots of publicity, no knowledge, you inevitably get into trouble, and then expect us professionals to come to your rescue.'

I've never heard that expressed by other professional seamen, whose attitude within my own experience has been to help and to help generously, but I've never forgotten what that one bastard said, and have been determined never to give his remark justification.

On 29 March I wrote obviously expecting to pick up the light on Rottnest that night:

> I was up till after midnight distilling water (so that after sighting land I would not have to linger), scrubbing out the saloon, mending the table, collecting ship's papers, and put a clean change of clothing ready. Feel ten years younger—God to be on land again—to sleep—I've not slept for more than two hours at a stretch for over eight weeks or something. Hell.

I sailed all the next day to the east, *Sheila* wallowing along with her barnacles, and after sleeping for two two-hour shifts in the evening I sailed all through that night. At times I could have sworn that I saw the loom of the light, and it might have been. No land was visible in the morning haze and out of it came a hard southeast by east, sending me far up to the north. After being very depressed I wrote:

> It is childish getting upset like this—I must be within 50 miles of the coast. All I need is the right wind and patience (not to mention food, drink and sleep!). But what sheer heaven to get

into port—to cable home 'all well'; cold drinks, hot drinks, sugar, bread and butter, fresh juicy tomatoes, oranges, apples, mutton chops, cauliflowers, and a hot bath, to soap and scrub, soap and soak and soak for ruddy hours—and a cold shower after, under gallons of water, and let it all run away down the drain 'cos there's plenty more. Clean clothes, sheets, and safe at night, and no more pumping out. Dear God, how can anyone living ashore ever be without content? May I never be disgruntled again while I have even some of these things. I'll be grateful if I get away with this and be unkind to no one, because gratitude to God is gratitude to all men—how true these things are, but how easily we cloak them when the danger is past.

Over the next three weeks I made three more attempts to make Fremantle. After each failure, having been blown north, I again made out from land and shipping to the south-west to recover south latitude, turned and ran in, and each time when within an estimated thirty or forty miles of Rottnest and land, breeze came hard from the south-east and pushed me too far north. It was disappointing.

One night I passed several ships, which seemed to confirm my proximity to Fremantle. One passed quite close, unaware of my existence unless she had me on her radar screen. I had a strong feeling she was the *Gothic* which I knew was leaving Fremantle about that time, bearing Her Majesty and the Duke away after their Australian tour. This was the third time during my voyage I had just missed the royal couple, and I thought what a hoo-ha the papers would make—'*Gothic* Saves Yacht in Mid-Ocean'. But in my darkness I let her pass, because neither *Sheila* nor myself were in a fit state to receive royalty.

I did call up two ships later, but neither answered. A third, which from the positioning of its lights looked like a tanker, replied far too fast for me to read (she should have sent to me at the same speed I morsed to her). I think I got the words 'Caltex' and 'Geelong', whereafter I told her that I was overdue and asked her to report me to Fremantle. I also asked five times for my longitude, which she may or may not have given—as soon as I began sending, each time she sent back something very fast without waiting for me to finish. Finally I nearly sent 'SOS', but by then was too angry to ask much help of anyone unwilling to give a little.

At this time I was becoming so weak and exhausted as to endanger the voyage far more than did the condition of the ship herself. The daily notes in my diary are fairly clearly written, but much of the time my mind was dazed and I lived in a trance, unable to grasp the seriousness of the situation; or rather it didn't seem to be serious—both the thought that I might be lost at sea or soon safe on land were incidentals my eternity lived outside. It was a state very akin to that I sought, but it held none of the vitality of that, none of the great wonder and joy. My very weakness, and the acceptance that I could do no more placed me in a state without desire; whereas the happiness I sought lay beyond it.

And so I felt perhaps like one of those conditioned humans portrayed in world-to-be fiction, who go on and on through each day and night without the will to grasp that by which we know happiness or unhappiness, yet vaguely aware that these exist and should be known; performing a routine of tasks without concern that they are never-ending, and without joy—the hours at the helm, changing sails, splitting wood, dressing my thumb, setting up the condensing plant, keeping the fire going, stowing the precious bottles of clean water, the patient and harrowing angling for dorado, the cutting into pieces, the cleaning up, the cooking, the eating into brief freedom from hunger pains, and the pumping, the never-ending pumping.

And so after the fourth failure to make Fremantle, and considering my growing weakness and the danger of fainting fits, I dared not turn south-west once again away from land and shipping. I decided to sail north-east with the winds until sighting the coast, and then sail up it to Geraldton. And so I threw belief away, but in my deepest being did not know whether that was submission or defeat.

All down the western coast of Australia, almost as far south as Fremantle, is a barrier reef of coral similar to but not so extensive as the Great Barrier Reef on the other coast. There are breaks in this reef through which shelter may be sought behind it, but it would have been dangerous to attempt these while *Sheila* was almost unmanageable. The nearest safe port of easy entry was Geraldton, 200 miles north of Fremantle.

My DR put me only just out of sight of land, but not being sure of this (of my longitude) I had to be very careful during the hours of darkness. If I was further offshore than my DR indicated, in running north-east I would run full tilt on to the Abrolhos Islands (The name is Dutch and means 'Beware'.) These are a low-lying, unlighted chain of sandy coral islands which run for sixty miles parallel with and about thirty miles off the main coast, which I planned to pick up and then run north inside the Abrolhos chain to Geraldton.

My navigation was undoubtedly affected by another attack of fish-poisoning that was just coming on at this time. After about a week on nothing but fish my body revolted, and my stomach (which usually only worked once a week on such a diet) became racked with pains and demanded unpleasant attention; the glands in my throat became swollen and ached, and a splitting headache throbbed with the slightest movement. All I could do for this was to stop eating fish for twenty-four hours, when the pain of sickness left me but that of hunger increased. (It was lucky I had always caught only just enough fish to keep me alive, otherwise this poisoning might have destroyed me.)

Noon sights over the next few days put me on the same latitude as the Abrolhos Islands, and as the mainland was not in sight I knew I must be outside them, that they lay between me and Geraldton. I did not dare to sail east during darkness therefore, and the only alternative was to lay hove to, losing time, increasing weakness, gaining nothing, drifting north. On 17 April there was still no land in sight and my sights put me north of Geraldton. As I had been unable to reach Fremantle from north of it, I could expect Geraldton to be equally unattainable. And the next harbour of any sort north of Geraldton was 300 miles away, Shark Bay, which would take a week I couldn't last.

The wind died that night and Sunday dawned over a lazy swell running under a glassy surface of dead calm.

On that day, Sunday, 18 April, I wrote: 'Fourteen weeks out today, over a quarter of a year drifting around half dead, entirely alone, it's fantastic.'

I lay in bed a little longer to luxuriate in the peace after so many days and nights of harsh winds. There was no need to go to the helm, nor to get breakfast because there was none to get, nor to pump out immediately because without the strain of movement the leak allowed me longer to rest. The peace brought a sense of quiet appreciation to a mind which had been living in a trance, and the written entry ends, 'Only God can now get me out of this, and if it is not His purpose to do so I've had it.' It was on this day that I finally accepted I was to die at sea.

There was no panic, no regret, and no resentment. I had set out for Fremantle against all common sense and seamanship, acting on what I took to be belief in getting there; and after the fourth failure to do so I had thrown away that belief with my decision to fly with the winds to safety. How could belief work if I threw it away, and whose choice had that been but my own?

I cleaned up the galley and tidied the cabin, bathed all over in the last of the fresh water, changed into a clean sarong, and took a noon sight as a matter of routine. During the afternoon I shaved, and with all tasks fulfilled I lay naked in the sun. There was no fear of death—I simply thought that at dark I'd go below and fall into a deep sleep, and perhaps only awake for the few moments before *Sheila* sank. There was the dinghy, but I preferred to stay with her.

The glowing red ball hurt my eyes as it sank into the sea; I stared straight into it as *Sheila* rose on each big swell, to see the very last of this last sunset I would ever see. I noticed jagged edges on the horizon where it cut the sun's sinking face, and blinked my eyes to clear the distortion, but each time *Sheila* rose they came again. So I climbed on to the boom to get more height, when the jagged edges looked like the tops of trees, and I remembered that the noon sight that day had put me on the same latitude as the northernmost of the Abrolhos Islands. I must have passed the outlying reefs very close during the night; Geraldton lay seventy miles to the south-east and the main coast only about thirty to the east.

Before wind came that night I caught another dorado, although none had appeared during the day. And after a meal I sailed until midnight, when I was so exhausted I was weeping with the agony of it and with the desperate desire to live now that land was so near. In the morning the wind was southerly again, precluding any chance of making Geraldton, so *Sheila* lumbered her way east towards the main coast which showed up just before dark. I continued inshore until midnight, when I feared the coastal reefs so turned about, hove to and slept.

The low coastline rose out of the haze next morning, 20 April, and I determined to get ashore at any risk. If there was no entrance through the barrier coral I would try to anchor *Sheila*, get ashore in the dinghy, and walk inland to phone for a tow from Geraldton—and if necessary sell *Sheila* later to pay for it. That would be better than deserting or sinking her.

Huts were visible ashore by noon, and also several fishing launches at anchor inside the reef. I could not see the entrance as the heavy swell broke in one long line of seething white, and so I planned to call out one of the launches to tow *Sheila* to safety inside—she manoeuvred too badly to attempt a narrow entrance even if it had been discernible.

Unlike Indonesian waters the sea-bed shoaled outside the reef, so I anchored in five fathoms on thirty fathoms of chain; in moments the heavy surge found the weakest link and the chain broke. I put down the second anchor on sixty fathoms of wire cable, and this held. For the next three hours I watched people walking among the houses ashore and along the beach, while I flew distress signals, fired off about fifty rounds of ammunition in groups of three, and sent volumes of black smoke into the wind by burning oil-soaked rags in a bucket on deck. By 4 p.m. I knew my only hope of help lay in going and getting it myself; it was doubtful whether the light dinghy could ride the surf over the reef, but what alternative was there?

The smooth rollers built up over the thousands of miles from Africa were no menace before they hit the reef. I took the dinghy slowly towards this and waited for three outsize rollers to pass, and then moved in closer and took the next smaller roller just before it broke—then spun the dinghy and rowed with all my strength.

I could see the jagged coral a few feet beneath me through the foam-flecked water, and astern the next wave came rolling in, rising higher and higher as it approached the outer edge of the reef, mounting into a great glass-like wall. I prayed for it to break. Slowly it became top-heavy, the crest curled, and then the whole mass crashed on to the reef and rushed towards me in a seething hissing wall. I swung the dinghy to face the broken water—the light craft stood upright on its stern, the bow high in the air, hovered for one tense second, then fell flat, and I struggled to get control as the violent eddies and suctions twisted and spun us like a piece of driftwood. I crossed the reef and the next wave was spent by the time it overtook me.

The wind and current over the next half-mile swept me far down the beach; there I was so exhausted I misjudged the surf and the dinghy was lifted end-for-end like looping the loop. I can remember the whole world was green, the light greenness of the water around me stretching into distances of darker green, seeing the sandy bottom under my feet, the moment's panic which gave my back and legs the strength to force away the dinghy holding me down. I flopped about on the surface out of my depth, until a wave picked me up and with a swoosh sent me far up the sand, and I dug in my toes and fingers while the water drained away and left me. My first land for exactly one hundred days.

My one idea was to save *Sheila* before she broke adrift from the second anchor. The dinghy's long painter was swirling in the shallow surf, so I caught hold of it, dragged the dinghy up, righted it, and tied it to a bush.

It was a strange feeling walking along the deserted beach, the struggle up the sand-hill to the nearest hut, seeing a girl inside drying dishes and noting the startled look on her face. She called her husband, a fisherman took us out through the reef in his launch and we towed *Sheila* inside. Here I was handed over to the wives for food and dry clothing while the men kindly tied *Sheila* to a mooring—the only time I have ever neglected to do this myself.

The men went back down the bay in the launch to collect my dinghy. They returned just after dark to say that the strong tide and surf had swept the launch on to the beach, and that she was by then a total loss.

31

*Illegal entry—'You're in Australia now'—move to Geraldton—
the use of Christian names*

Far back in the book I wrote that for the first few days of each voyage I was on edge and unsettled until the way of life on land departed and I merged again into the life at sea; so it happened after arrival from the sea. The thoughts and ideas which were so real and practical there clung to me for my first days on land, and made me very vulnerable.

The girl and her husband whose cottage I had first entered were kindness itself. They gave me clean dry clothes to wear, and she prepared a delicious meal which I could barely touch. Longing as I did for food, so much revolted me, and I was to learn that for months I could eat only a very little at one time, because my stomach had shrunk and lost its elasticity. But my thirst was raging and unquenchable, and even more ravening was my desire for sugar, because I had had nothing sweet for over eight weeks. The girl made me jug after jug of hot sweet coffee, which I drank in a frenzy of desire I could not control.

And then I went to bed, a soft, warm, dry bed between clean sheets, with freshly ironed pyjamas against my skin, with no danger, no need to awaken within the hour, no seconds of sudden panic in darkness as the sea sloshed against my bunk, no pumping out. And *Sheila* was safe at her mooring, where at rest and in quiet water she leaked little. It was the fisherman's own mooring, the weight and strength of which he had described and guaranteed.

In England there is an unwritten law that whenever you occupy another's mooring you ask when you will be required to vacate it, so as to cause no inconvenience to the owner. So that very first evening I asked the fisherman, 'When will you want me off your mooring?'

'You can have it as long as you like, but when you do leave ensure the float is OK; I'll pick it up sometime.' (I had lost my second anchor also in the coral outside the reef, which left me with no other.)

Next morning my host drove me into the nearby country town of Northampton, where I immediately reported to the police and other officials, sent cables home, and contacted my bank in Perth. There was some excitement because I had long been reported overdue, and because Horrock's Beach where I had landed was not equipped to receive ships from abroad. Such ships must enter the country of arrival at what is known as a First Port of Entry, where there are the necessary officials to grant it clearance-in—the Customs, Health, Immigration, and so on.

That very day the Customs Officer arrived from Geraldton, in a delighted state of agitation at this entirely unusual event in his normal routine, and fortunately for me he was the ideal official to deal with such an occasion.

Harry knew his regulations thoroughly (and Australia had more of these than any other country I visited) and had the strength of character to apply them, the flexibility of mind to fit this unusual event to them, and a pleasant personality which made co-operation easy for me. I had nothing of value except the ship herself, the statuette of Kwan Yin, my Leica and cine cameras, and a .303 rifle; the rest of the cupboard was bare.

After a couple of days I moved into a shack, and while resting began an article covering the voyage to bring in some much-needed capital. Several days later the fisherman came to see me, and to my amazement his manner was grossly hostile.

'I've come back to pick up my mooring so you'll have to get off it. You're holding me up now, so get moving.'

As he had previously said that he was in no hurry for the mooring, I had planned to stay several weeks at Horrock's Beach, to recuperate before attempting to take *Sheila* on to Geraldton. This was a dread I pushed out of mind until I felt stronger, because as I had failed to make Fremantle and Geraldton, there was every likelihood that after leaving the shelter of the reef I might again be swept north; I explained to the fisherman that because of his own first statement I had made no alternative arrangements, that I had no anchor nor the money to buy another.

'That's your pigeon. You're in Australia now, boy, where every man helps himself. You've been here a week and all you've done is sit on your bottom writing about it. I'm taking my truck back to Geraldton now, returning tomorrow, and if your bloody boat isn't off the mooring by then I'll throw it off.'

'Then you'd better give me a lift into Geraldton,' I said. I had no idea what I was going to do, but there were anchors and new mooring rope in Geraldton and somehow I would have to get them.

On the way through Northampton I called at the bank to see if the Perth branch had forwarded any mail. There was one letter only at that time, and it contained £20 for an article sent from Java. Before arrival in Geraldton the fisherman said, 'By the way this lift will cost you a quid.' And when I got out later I asked if he would give me and my heavy gear a lift back with him the next day, but he said it would be inconvenient.

Harry was the only person I knew in Geraldton, so I went to ask his advice as to where to buy my needs. He took me around to Tropical Traders in his car, and I bought a fifty-six-pound anchor and thirty fathoms of new 3in. rope on credit guaranteed by him, because he realised I would need the £20, it being all the cash I had. That night he put me up in his home, and next day drove me and my gear to Horrock's Beach. These are incidents of my voyage I do not forget.

One look at the harbour told me that it was too rough to change *Sheila*'s mooring that afternoon. The hard southerly flew over the reef piling the big ocean swell into the anchorage, and the extra water could only escape back to the sea by the one narrow entrance through the reefs. This caused a four-knot tide and rough water. My plan was to take out an old tractor wheel which lay on the beach and use that as a mooring on *Sheila*'s chain, together with the new anchor and rope—I was not prepared to leave her at only one security in those conditions.

'You call yourself a seaman,' said the fisherman contemptuously, 'and can't even lay out a mooring.'

'I know exactly what my dinghy will stand and not stand,' I replied, 'and to take it out heavily laden in that sea will sink it. The whole operation will be just a waste of time. I suggest we wait until tomorrow morning, before the daily southerly starts.' I did not say in retort to his remark about my seamanship that I had never lost a power boat on an open beach, because fortunately I never thought of that crushing answer until later.

The fisherman's hostility came from the strange attitude of some Colonials who have never left their own country, who consider that they are the only truly individualistic people in the world, that they only have the courage to speak their minds, or in other words to be thoroughly objectionable. They only have self-reliance and toughness, and accordingly view all others with contempt. I understood this attitude well because I had been exactly the same at the time I went to Sandhurst, but unfortunately for me the fisherman had not had the levelling experience of that aristocratic establishment. My only course now was to prove myself right by demonstrating its failure.

It had not been my fault that the fisherman had lost his launch, but the fact remained that he had done so while helping me. It had been insured for total loss only, and as the engine was recoverable this waived any claim. Some days later he told me that this had been confirmed and he was to get nothing.

The launch was his main capital asset and his means of livelihood with which to support a wife and family. After a lot of thought I offered to sell *Sheila* and pay him half the proceeds, which was not the cost of his launch but it would be enough to help.

'No,' said the fisherman, 'I don't want you to lose your boat, but I think you should pay me something for towing you in.'

I worked out that the most anyone could have possibly claimed for the mile tow was £50, so because he was getting no insurance I doubled it, but with the warning that he would have to take it in dribs and drabs as I earned it. His start of surprise at the voluntary mention of £100 told me this was far more than he had expected; but my assessment was not made while comfortably ashore, it was made by going back in memory to that dread evening when his launch had arrived alongside *Sheila* to tow her into safety. It was worth £100 to me then.

Harry and a friend of his helped me load the heavy wheel and chain into the dinghy, and then by wading I towed it up the beach about 500 yards, uptide from *Sheila*. My plan was to row straight out from the shore, and by the time I reached a line ahead of *Sheila* the current would have swept me down to where the mooring had to be dropped.

The waves slopped into the laden dinghy lowering its freeboard even more, and long before I reached the dropping zone it sank with the heavy wheel still in it. I was too weak to swim to the shore and the strong tide would have taken me out through the entrance, so clinging to an oar I made certain I kept straight uptide from *Sheila*; and as the tide swept me to her I caught her mooring chain and clambered up the bobstay.

The same evening that the dinghy sank Harry had taken me aside.

'You know that I'm an official but am now talking as a friend, although I must warn you that I will have to report your answer. That seems silly, but I thought you might prefer to answer me than the police.'

I was still wet and shivering from my dip, and had no idea what he was talking about.

'Two adults and three children have been to the police and stated that they saw two people on board *Sheila* before you came ashore. As only you got off later, they think that the other must have hidden and landed after dark. Did you land an illegal immigrant?'

'I can only deny it, Harry,' I said, and left it at that. I was rapidly losing patience with Australia, and this was a very serious allegation. I lay awake a long time that night thinking about it, but had the answer by the time the police contacted me next morning. They asked me bluntly if I had landed an illegal immigrant.

'My denial against the evidence of five others,' I said, 'probably means nothing to you. I ask you to bring a formal charge against me immediately, and let me have those five people under oath in a court of law, and I'll give you my answer out of their mouths.'

'They won't make the statement under oath,' said the policeman.

'Well, either you bring a charge at once, or tell me now that the matter is ended. Which is it?'

The police decided to drop the matter, and I never heard anything more about it officially. The rumour did however circulate and the stigma I think initially did me harm. The Australians are terrified of illegal landings on their long, deserted stretches of coastline. My defence was that the main accuser had been on board *Sheila* from the time she was taken in tow until I was put ashore, and after that until he left her at her mooring. If he had seen two men on board, and if he saw only me get ashore, why did he not ask me where the other was or search the ship after I had left her?

Sheila stayed at her new mooring beside the old, which was left there for another fortnight. I dreaded going back to sea, but she was not safe in that anchorage with the storms that could blow over the reef; and I had to get a job, which was not possible at Horrock's Beach. I put three weeks' supplies on board, just in case I was swept north back to Shark Bay, and tackled the problem of cleaning-off below the waterline.

There was too big a surge on the hard beach for me to prop *Sheila* up and clean her off between tides. So I slewed her as close to the shallow shore as I could, about forty yards off, and tying a rope round my waist against the strong current I dived overboard with a scraper. This proved to be useless, as the tide spun me like a spinner and any effort against the hull simply pushed me away from it. I was terrified by the thought of sharks, as these had been seen inside the reef and I later confirmed it for myself; yet it was no use asking anyone to stand watch, because what could they have done had a shark come swiftly while I was under water?

So one morning three weeks after arrival at Horrock's Beach I took *Sheila* out through the reef in the same condition as she had arrived, and headed for Geraldton about thirty miles to the south. My nerves were in shreds, and that was one of the hardest things I have ever done.

The winds were kinder than they might have been and next evening I anchored just outside Geraldton, because the wind blew straight out of the harbour and the engine was out of action. The following morning a young fisherman offered to tow me inside, and when I was too weak to get the anchor in he leapt on board grinning and got it in for me.

Perhaps the hardest part of a voyage like this is the end of it. I longed to get away somewhere quiet, away from officialdom, particularly from all responsibility, to rest and sleep, to find a friendship with understanding but without the never-ending questions, some of which were of a very personal nature. But there were no Kitty and Fien in Geraldton. This attitude of mine was of course lack of guts but I was wrecked, destroyed, because having given everything I had nothing left. No one of course realised that.

My recovery from that voyage was to take far longer than even I realised then (in panic I put it down to old age!). It took a whole year to recover physically, and during that time the thought of going to sea again was a dread I only endured by banishing the thought whenever it came. I knew I could not give up but I did hope that the matter would be taken out of my hands, as by *Sheila* breaking free from her moorings in a storm and being wrecked, as some other boats were during that year. I took every precaution to ensure she didn't because you can't fool yourself on these things, and so I left the matter to a greater power for decision. This was stupid, because the decision is always finally our own.

But *Sheila* had to be slipped, repaired, and put at a mooring safe enough to hold her for a year while I earned the small fortune required to refit and complete the voyage. After enquiring about local prices I estimated that I would need £600, which included the £100 to the fisherman. It was no use dreaming of getting home at that time.

I searched the city's rubbish dumps and the waste land behind the port, found two engine blocks, dragged them one at a time to the edge of the wharf, and as no one offered to help I rigged blocks and tackle and lowered them over the side to just under water. Then I took the weight on to *Sheila*, and with the engine revived she carried her own moorings to a place in the harbour where she could swing free of other craft.

She was later slipped, cleaned-off, painted, and the shipwright repaired the leak. He kindly undertook this task in anticipation of the time when I should have earned the cash to pay for it, although on such occasions *Sheila* herself is my security.

During this time I lived ashore in an old furniture van converted into a caravan, and began settling into the Australian way of life. This was not easy, because the last land in which I had been settled for any length of time had been England. The two peoples are basically identical as I have found all people to be, but they are as far apart superficially as they are geographically. It may be six months before you call an Englishman by his Christian name, but by the time this is allowed you probably have a friend for life. But when you meet a stranger in Australia he asks, 'What's your name?'

'Hayter,' I'd say.

'No, what's your Christian name?'

'Adrian.'

'OK, Aid, mine's Jim...' and you might never learn his surname. This made the way to friendship easier, but unless the mutual ingredients were already there and until they were known, the free use of Christian names did not mean a thing.

The general lack of formality, particularly, I suppose, after my life in the Army, brought several shocks. One incident happened at a small grocery store where I called most days for casual needs; it was run by a vivacious woman in her late thirties who always chattered with me for a few minutes. In the doorway there was often an old dog, which sat there with an air of friendly ownership and I usually gave it a pat as I left the shop.

On about my fourth visit I collected a loaf of bread, chatted for a time, then turned to go home as the woman busied herself at the till. At the door I stooped to pat the old dog and said, 'Bye bye, Poppet,'—and without looking up from the till the woman called cheerily,

'Ta ta, see you to-morrer.'

With *Sheila* on her mooring I began to look for a job.

32

*In the slaughter-house—the fencing contract—the shearing shed—
on the wharf—social life—a harbour bar*

There were plenty of jobs at that time, and these fell roughly into two categories. The basic wage jobs were easiest to get and almost unlimited, bringing about £12 a week after paying tax; the others were mainly on a share basis, such as well-boring, fishing, contract fencing, or skilled, such as shearing sheep. As much as £60 a week and sometimes more could be earned in these, but they held no easy entry for a stranger because he needed a mutually acceptable partner, some capital and equipment, or some skill; and at that time I had none of those things. But only in one of those jobs did my chances lie of earning enough to refit *Sheila* within a reasonable time. And time was a problem, because a ship deteriorates more quickly and suffers more damage at anchor in a crowded harbour than she does at sea. The longer I took to earn enough to refit, the greater the expense of re-fitting, the need to stay longer, and so on into a vicious circle.

In the basic wage jobs from my experience a full and reasonable eight-hour day was seldom done, and if the employer complained it was too easy to leave and work for another less exacting. It amply displayed to me the absurdity of a community basing its economy on an 'enlightened' system which guarantees a minimum wage without also guaranteeing a minimum output. Unfortunately it seems that the only power to impose a guaranteed minimum output is a degree of unemployment—and I'd like to hear one of our national leaders make that sensible statement and keep his seat!

Harry introduced me to Mick, who was an active figure in local commercial circles.

'Yes, Adrian,' said Mick, 'I'll give you a job. It's only on the basic wage but it will keep you going while you look for something better, and don't hesitate to let me know when you find it.'

Winter was then coming on, so each morning thereafter I rose before dawn, lit the lamp, and got my breakfast. This gave me time to pack my lunch and walk along the beach to the town to meet Jeff by 8 a.m. It also gave me time to wash up before leaving the caravan, because it is horrible to return home after a day's work to find the place in a mess, with no clean pots in which to get supper.

Jeff drove the concern's truck and my job was to be his off-sider, or in plain English, assistant. He was a tall, rangy individual who might have been considered rather rough in an Officers' Mess; he didn't give a damn for anybody, but inherently he was more of a gentleman than many who claimed that title.

Each day we drove several miles out to the slaughter-house and holding paddocks (Mick ran Geraldton's meat supply), where we yarded sheep and cattle, repaired fences, laid water pipelines, erected huge water tanks, cleared scrub, tore out bushes with a tractor, carted indescribable filth in huge buckets from the scene of carnage, and drove to farms far and near to collect animals for slaughter.

In those early days I wondered how Jeff would regard his off-sider, and so kept quiet to await developments. After a few days he took me in to meet his family as we passed in the truck; he had a contented wife and five happy children, which was a greater achievement than mine. And that day the others at the slaughterhouse ragged him.

'Hell, what's the boss thinking of, giving a no-hoper like you an off-sider?'

'He's not my off-sider,' said Jeff. 'He's my mate.'

We worked as equals but we were not equal. Most of the jobs were sheer physical strength, and at times I could not do my fair share nor anything like it. I was still about a stone underweight and my endurance was nil, but Jeff did his own share and the surplus of mine as well, and neither chided me nor ever showed resentment.

Every Saturday morning we went to the office and Mick gave us an envelope containing our week's wages. On the outside was marked the full wage, any additions such as overtime, any deductions such as tax or refund of advance received; and the balance co-ordinated with the cash inside. All this I took to be quite normal, implying the same care with which we had always paid out our men in the Army; but in this respect that first job of many others was unique.

Sundays I looked forward to as never before. I slept in until about nine o'clock, cleaned out the 'house', washed dirty clothes (and they were dirty), visited *Sheila* and checked her moorings, and caught up with my writing. Without those days of rest I don't think I could have continued at that time. It was not only the early rising and the return home filthy dirty, with just enough daylight left to have a cold shower and change (there was no hot water system in the camp); it was a lonely period, partly because of being run down physically, and partly because the land conflicted with the sea which still held me.

After working for Mick for three months I heard of a fencing contract far inland, so went to collect my account; but the credit I received was greater than that expected.

'What's this extra for, Mick?'

'You're entitled to a day off a fortnight, and as you worked those days you get overtime.' He needn't have mentioned that because I'd never heard of it, nor did I in any of my other jobs. If all employers were like Mick, unions would not be necessary.

I had no truck, equipment, nor mate, which are the usual requirements for a fencing contract, but I thought the job was worth investigating. I had no idea how long it would take me to erect a mile of fencing and made enquiries about the normal payment, which depends mainly on the type of fence and the kind of ground over which it is to run. You can dig a post-hole in sand in a few minutes but in hard clay, sand-stone, or solid rock, it of course takes much longer and time is the essence of the contract.

The information as to the normal payment varied widely. If I asked an employer during a visit to a local farm, he told me about £40 a mile over reasonable soil; if I asked labour, they said as high as £120 a mile. I also made enquiries about my potential boss and found that he was a stout pillar of his church, and a big influence in the councils of his district, so I assumed that his offer would be reasonable—apart from morality, if a man gets a bad name as an employer he only works against himself, particularly far inland where labour was hard to get.

The boss offered me £33 a mile, because he was to dump the materials on the line, give me a shack to live in, supply all the crowbars, shovels, pliers, hammers, axes, plus a horse and cart for me to move gear on the site. He drove me over the rolling country where the line was to run and I asked, 'What's the soil like?'

'It's easy,' he said. 'It's all sandy.'

The old fence had to be pulled down, the posts stacked, the wire rolled up and the barb coiled and stacked separately. We eventually agreed on £45 a mile, but I wanted extra for pulling down the old fence which would take a lot of time.

'We'll come to some arrangement about that,' he said.

'No, let's settle it now,' I demurred, 'then there'll be no confusion later.'

'Don't you worry,' he said benignly, 'I'll see you right on that.' So I dropped the matter and it was agreed I would begin a week later.

Saturday came, and just after dawn I loaded my gear into the truck plus a week's supplies, and because I had learnt more of the fisherman's advice ('You're in Australia now, boy—'), my Primus stove and two gallons of water. Then we set out for the great beyond.

The boss met me at the bach, a corrugated iron affair with a falling tin chimney; the doors had broken latches, it was filthy inside, the kitchen walls were black with soot and grime, windows were smashed, and there was not a stick of furniture in it, not a table, bed, or even an old broom.

'I wonder if there's any water?' mused the boss. It was a question he had already had a week to consider. The big empty tank open at the top held a mere foot of slime, the last resting place of a decomposed crow.

'I'll do something about water for you,' said the boss, 'but now I'm off for tea. Be back this evening.'

So I opened a tin of bully, lit the Primus for a cup of tea, and sat uselessly on my bottom for the rest of the day. It was no use unpacking until the place was at least swept, but there was nothing to sweep it with. I was seething, and decided that if the boss didn't return with at least a broom, I would go back to Geraldton in the morning.

This small incident discoloured the relationship from the very beginning and other subsequent incidents, small and perhaps nothing in themselves, were biased adversely by it. No officer in the Army who neglected these 'small' things in relation to his men ever got the best out of them, backed as he was by the full power of the Army Act. With further experience I became convinced that the antipathy between capital and labour is not basically one of wages and attachments, but of something more subtle and difficult to define; not all the rules and regulations, arbitration decisions nor the demands conceded to unions can ever grasp it. Perhaps it's simply human consideration and nothing more—by both sides.

The boss took me out on the Monday morning and showed me the type of fence he required. Next morning I was up before dawn, had breakfast, caught the horse and led it to the open shed, where rested the heavy dray and a tangled mass of harness. Somehow I had to connect these things together to make a unit of transport, and used the process of elimination learnt in geometry at school—I knew the blinkers, bit and bridle went on the front end; a long, heavy strap ending in a dung-stained loop obviously went aft; and the collar with two curved iron spikes with rings on them could only go somewhere round the neck because they wouldn't go round the girth.

Fortunately the old mare was quiet and patient, but several times she looked round her shoulder in amazed surprise at what was going on. She reminded me of an elderly aunt in England who used to look at me over her spectacles whenever I made a remark that might have been frivolous. Thereafter I found that driving a dray was far more difficult than steering a ship, and at the first departure the hub of the wheel caught the corner post of the shed and left the roof forlornly sagging.

There were heavy frosts in the mornings and the nights were bitterly cold, but at first the days were clear and beautiful and the hard work was a joy, because I found that it contained all the needs that I sought from my lone sea voyage, and was an equally certain path towards attaining it; so long as that came, with the increasing contentment its approach brought, I had no care of how, why, when, or where it came. But I soon found that in spite of the long hours the contract was going to bring me little more than the basic wage. I had been deceived.

As the fence-line rose up the slope of the hillside the ground became more stony and harder until at the top a strainer hole took three solid hours with crow-bar and shovel. These conditions in themselves did not matter, but being so different from the sandy soil on which the contract was based they destroyed the sheer happiness I had found in the work and my whole reason for living. I am incapable of excluding resentment when my given trust is exploited, and resentment or the fight to exclude it denies the possibility of contentment.

Later when I recounted the story to others in Geraldton they laughed and asked, 'But when the boss said it was sandy soil, didn't you take a shovel and test along the line?'

'No,' I said, 'of course not. That's the equivalent of calling a man a liar and how can you work with him after that?'

'Stick around, chum. You'll grow up.' That's exactly what I was fighting not to do.

During those three weeks I needed money to pay store bills and remittances to the fisherman. I asked for payment of the work completed, and three times I asked for a settlement concerning the pulling down of the old fence. Each time I was told not to worry until I was finally given a cheque for £50, not for a mile of fencing but 'To advance'. This did not confirm the rate of the contract, so I pressed again for a decision about the old fence.

'Oh, that's all in the contract of £45 a mile,' said the boss. I hate arguing about money and at first thought this reversal was forgetfulness of a verbal contract, but how could it have been when on three separate occasions I had been told not to worry, and on not one of those occasions had I been told that the old fence was included in the £45?

I struggled on for a time, because although the earnings were little above the basic wage, the savings were greater without rent to pay. Then came days of rain, the cuts on my hands festered and the pain slowed up the work, and in my bitterness these all became excuses for not doing better. 'Why sweat out my guts for £45 for a bastard who should have been paying £80 or more? With a labourer's work you take a labourer's mentality whatever your background, and after this I could so easily understand not only the developed power of the unions, but the appeal of communism itself. And the fault is not the labourer's.

The flies were unbelievable. These found great delight in the sweat on my face and the various excretions which ooze in a human's ears, nose and eyes. Both my hands were occupied with my task, but a shake of the head or a swift brush of the hand had no effect on these flies—they died on the job, and had to be crushed in whichever organ they happened to be interested.

A larger variety laid maggots faster than any previously encountered, and at lunch time no matter how swiftly I took a chop from the packet those inside were always blown. These flies didn't have to settle, they employed an aerial top-dressing technique and could only have sprayed their loathsome progeny. I always brushed off what maggots I could, but it became a choice of maggots and chops, or no maggots, no chops.

Then the station shearing team didn't turn up to crutch the flock. The boss managed to get a neighbour and two old sweats who had spent most of their lives in shearing sheds. He asked me to make up the fourth stand at the recognised rate of £2. 10s. 0d. a hundred. I had hardly seen a sheep since going to Sandhurst, but as the fencing contract was so unproductive I agreed to try my hand.

The two older shearers were helpful and often amusing with their somewhat cynical humour. One day they were discussing a mutual friend who had evidently been quite a character and not overfond of hard work.

'He's been crook for some time, hasn't he?' asked one.

'Yeah, haven't you heard? He died a week ago.'

'Go on? That's the first useful thing he ever did.'

But when we went on to shearing I had to give up, because with my lack of skill and aching back it was simply not an economic proposition. So I was given the job of shed-hand, which consists in keeping the shearers' pens full, picking up the fleeces, sorting the various grades of wool, and sweeping up the odd bits and pieces from the shearers' boards. It is not a hard job but it is continuous bending, lifting, and never a still moment. The boss said he would pay the usual rate, which I found was £15 a week for a boy under eighteen and £21 for a man.

At the end of shearing I asked for a final settlement, as it was time for me to return to *Sheila* and prepare her for my onward voyage. My whole account since the fencing was clearly made out, tax deducted and receipted.

'Only two things,' I said. 'You've paid me for shed-hand at a boy's rate, and I'm over eighteen.'

'Yes, but it's only a boy's work isn't it? What's the other?'

'You've not deducted my keep while I fed in the homestead during shearing.'

'That's all right, we can forget it.'

That family had been friendly and good to me and I had fed with them as their guest, but how could I feel gratitude after the deception of the sandy soil, the old fence, and the arbitrary payment of a boy's rate? The local church had a shaky prop.

On return to *Sheila* I found that the shipwright had not done much of the work which out of his own mouth he had accepted. And as for the sails I had ordered locally—the storm jib was an excellent sack, the mizzen was three feet too long in the hoist and the main three feet too short. In the moments following my discovery of these details I became as excitable as the New Australian sailmaker. It was no use demanding that he alter them, because I had neither the time nor the confidence that he could effect an improvement.

'She good,' he asserted. 'I make sails for all fishermen, only you complain.'

Not one single job I ever had done in Geraldton was anything but expensive, and the delay in time cost me more than the drain on my hard-earned wages. The inefficiency was not deliberate nor malicious, and all concerned (except the excitable sailmaker) were quite charming about it.

'Aw hell, she's right. If she don't come good bring her back and I'll have another go.' This attitude may not be due to the hot climate just below the tropics, because I had also observed this charming outlook towards time and obligation in the cold mists of southern Ireland. Both wonderful places for a holiday.

I had ordered new running and standing rigging for *Sheila* and had decided to do all the wire splicing myself, as that of experts had twice proved to be unreliable in the past. I don't know why splicing is usually considered to be so difficult (which was the reason why I had not done it myself previously), for it can be learnt in a few minutes. There are about fifty wire splices in *Sheila*'s rigging and each was packed with grease, parcelled, served, and given three coats of thin tar. All wire was given three coats of boiled linseed oil, in my experience the best protection for standing rigging. All blocks were checked, greased, the wire strops renewed, and the masts and spars were all rubbed down to bare wood, treated with oil, dried and rubbed again, then given three coats of varnish. *Sheila* was slipped and anti-fouled, the masts re-stepped, and she was ready for sea. I sailed her on excursions outside the harbour to stretch her baggy mainsail and to test for the leak, which appeared to have been satisfactorily repaired.

My savings were not enough, and so I arranged ahead for a job in a Fremantle wool store at three guineas a day, with which to buy supplies to take me on to New Zealand. And for two weeks while waiting for *Sheila*'s engine to be repaired I worked on the wharf.

The wages were high and the work controlled to lightness, but the nebulous air of unco-ordinated effort let me walk home each evening with not a jot of satisfaction in the day's work done. That sounds smug, but anyone can test it for themselves, and no wages can compensate for it; much later when I again had to find work I never returned to the wharf.

During those months before *Sheila* was ready I lived in a boarding house, because living on board a ship in the process of refitting costs more in morale than the rent ashore costs in cash. The long evenings were a problem, but there is a point to mention before I deal with that.

After visiting many ports I had come to make my initial assessment of each new place on two simple observations—the ties men wore, and the state of the public lavatories, which I was forced to use when having no abode ashore. I found without exception that the gaudier the local ties the filthier the artistry on the lavatory walls; where the ties were more subdued the walls were clean. The taste in men's ties proved to be a fairly accurate assessment of general local attitudes in all things.

There were dances twice a week when you simply paid 2/- entrance fee, but I wasn't happy about my clothes because I had economised on these. The local youths wore suits mostly cut to the American style, or comfortable slacks and an open shirt. The girls were almost without exception very well turned out, and never have I seen such a high concentration of such sheerly beautiful girls as in Western Australia. It amazed me the casual treatment they received from the men, girls who in London would have walked over the bent heads of princes. Anyway introductions were not needed, you simply asked anyone you liked to dance, whereon she might refuse, or accept, or let you take her home; your personal attractions were greatly enhanced by the ownership of a car.

An alternative evening entertainment was the cinema, which was always a double-feature programme. In almost all cases the first feature was some version of cowboys and Indians, which I found sometimes made the blood run faster but left the intellect undisturbed. And 96 per cent of the second films depicted some form of American racketeer or gang activity, where a bad boy met a bad girl, a spark unaccountably glowed between them, grew into a love to save them from the Chair, and enabled the audience to assume subsequent years of marital bliss. So easy.

It has always amazed me why the Americans turn out so many of these films, the widest portrayal to the outside world of what life in the USA may be assumed to be. Their police are almost always depicted as giant physical oafs, without intelligence or compassion but brutal and corrupt, and the last to whom those in trouble go for help, which is hardly surprising. The overall impression is that the average decent American citizen lives in a constant state of fear within a jungle of evil activity, until a dispassionate observer can only conclude that this great nation is the most barbaric in the world today.

And so the cinema offered little consolation.

The last remaining and most popular male pastime was drinking. Europeans in the East have earned a reputation for heavy drinking, but there it is mostly over leisurely hours after dark and often interrupted by dancing, cards, billiards or even conversation. But in Geraldton it was hard and fast, regardless of brand or mixture, resentful of interruption, and began at speed before the sun had set.

The pubs closed at 9 p.m. in Geraldton, but the attitude towards drink was born of the days of six o'clock closing, when a man knocked off work at five and had an hour in which to slake his thirst. In the public bars it was still a fight to get a drink, a feat to retain it, and a discomfort to imbibe it; and so more drunks will be seen at closing time in a small town where hours are controlled to limit the evil of drink than will be seen over months in Paris where drink is available for twenty-four hours a day. There is no evil in drink itself but only in the way it is used, and that demands an education which allows free opportunity to learn its blessings, and severe penalties to clarify its evils.

In the long evenings I usually read, or made out lists of gear and stores needed for *Sheila*, trying to estimate costs against my earnings to ensure not over-committing myself; or wrote notes for my book; and there were letters to friends in my regiment or of days past, or to others made en route, and to my home which seemed to bring it to me where I had no home. But some evenings a restlessness drove me outside to walk for hours along the darkened wharves, or to the pubs—not to drink into escape, but to fill this vacancy inside me. It was not a desire to be neutralised, but a need to be fulfilled.

I entered the hot crowded room, and made a way between the scattered tables and chairs haphazard among the noise, through the reek of spilt beer mingled with the smoke. I ordered a Drambuie, a drink in harmony with my mood, and found a table against the wall with a wide view of the whole room.

In the far corner, immediately noticeable because of their isolation, was a youth in a blue suit and a floral tie, obviously his best, and with him sat a slim girl. They looked self-conscious, their new love at discord with the slopping beer, the raucous voices, and the dreadful jagged piano. The evening they had dressed so carefully to enjoy now bewildered and frightened them.

No one seemed to notice me; I sat invisible, looking through the reek of foul air at a room of livid faces, urgent with excitement or slack with approaching nausea, mouthing their words silently because I could hear none above the noise they made. They were unguarded among their own kind, allowing me to look deeply into them. A lumper's face showed all that I have written of what that kind of work does to a man; a few fishermen hardened by the sea were sodden now by the very richness the sea allowed them.

The younger women there, except the slim girl, were all setting out on the road leading to what the older women already were, women who had grown up with little respect from men and could not now incite it, married young to a husband who already drank heavily because it was hard to find one who didn't—the days of the engagement when she had thought her love would reprieve were long since forgotten, and with despair the evil had embraced her also.

Of the older women some were widows, some divorcees, and some just going their own way while husbands slept it off after a three or four day blind. Some were skittish, hoping a younger man might be drunk enough to give them what they still wanted; all were prepared to tolerate the sodden old fools with the money to buy more beer.

Everyone in that room wanted the very thing they killed—it was seen in the departure of the young couple, their disappearance into the night to find it in the touch of their hands, their vulnerability as yet unlost, saving them. It was seen at times in a fleeting gesture of affection by a drunken male to a woman beside him, fleeting because its impelling need broke briefly through the beer-laden barrier and was quickly banished, banished before it could be rejected or ridiculed; because if that is shown to another and ridiculed it brings perhaps the deepest hurt we humans can know, by which we learn its value and feel its terror.

They drank to forget just that, which the women had lost because men had either flung it back or shown no need of it; the men suppressed the need because they feared its ties, or saw no hope of its being fulfilled, or because being one of the boys they had strangled it in its youth—a callous scuffle in the sand-hills, the return to the pub to boast of it, in loyalty to their friends labelling the girl to further enterprise. And so now they drank with frenzy to forget what they no longer knew—the shouts for more beer, discordant voices raised in song, the clanging piano, the whole room filled with smoke, and reek, and noise; everyone having a good time.

Dear God, I was sick of it—sick and sick and sick of it. It was travel to far places, it was life, it was how people lived; it was written in envious books to carry home-loving people into worlds they never saw, but I could not escape from it—I had seen it in the Mediterranean ports, in the Middle East and Asia, with white men and brown women, with brown men and white women, with cheap *vin rouge*, or smoky arak, or gin, or beer; it didn't matter, none of those outward differences changed the same basic, the same monotonous, the same pitiful tragedy—the callous defence to escape its pain denying the beauty of its reward.

I thought of the days that were past in London, Singapore, Paris; the swift cars, the beaches and the rivers, the cleanliness and the comfortable clothes, dancing, concerts, returning to a beautiful home lazy and content, wine-taken into a kind philosophical mood, sitting on the floor with friends who lingered late into the early hours—talking of books, music, Paris and Bali, of love and ideals.

That was culture in the higher income bracket, but this was barbarity in the red. The colour springs not from hard conditions nor envy of others, but from a freely accepted dictatorship which gives no outlet to the intelligence and sensitivities lying dormant in us all.

The lights flicked off for a second, a moment's hush, then the roar for last rounds and bottles to take to a home they no longer had. The great barrel of a body eased off the piano stool and rejoined her party, demanding her liquid reward. Two blue-shirted policemen, alone as policemen always are alone when popular law halts popular pleasure, stood unsmiling in the doorway. The lights flicked again, and the shrill refusals of the barmaids pierced the slurred demands of those not yet satisfied. Some moved to the door, others followed.

I walked home alone, wondering why we humans destroy ourselves with the very things which can give us happiness, knowing I myself had done the same in the past and would do so again in the future; but I did not know then that in the future lay a person who would lift the burden.

33

Departure for Fremantle—the return—builder's labourer in the North-west—a 400,000 acre sheep station—hospital

The full description of my last days in Geraldton, planned to close the last chapter, now seems unimportant and even irrelevant. But one chance meeting was important in the early days of refitting, when a young farmer joined me on Sundays from his week's work to help me on board *Sheila*. I think he respected my venture, and I grew to respect him and his family because of the example they displayed in their mutual relationship to one another. They gave me what I sought for myself, and it would only be an impertinence to say more.

It eventually came about that Bob sailed from Geraldton with me as far as Perth, where after a few days of associated frolic he was to return home by land.

I was not happy leaving Geraldton because of the debts left behind me. It said much for the people concerned that they accepted my assurance that I would earn the sums required in Perth, and send them before sailing for New Zealand. The summer southerlies had already started but had not then reached full force, and it was important to cover the 200-mile stretch to Fremantle before they did. But no one knows the future, and leaving debts behind me brought an uneasiness out of all proportion to the amounts concerned.

The days passed pleasantly but the head winds and seas begrudged our passage south. *Sheila*'s hull was clean and freshly anti-fouled, but with the cut of her new mainsail she only made six points (68 degrees) off the wind instead of her usual four, which of course increased the distance to be sailed.

Bob had told me that he might be sick, and was, but having determined not to disrupt my normal single-handed bliss he managed his affliction cheerfully. In anticipation however and without saying anything to him, I had bought a baby-food which claimed to be gentle, nourishing, and easily digested, which I thought would be acceptable and beneficial to a sick passenger. Never will I forget the day when I cooked up a plate of this baby-food, added milk and sugar with care, and handed it to this tough male in his distress. The mixing of this gentle food had been difficult and there were some rather slimy-looking lumps in it, but my gesture was well-intentioned—the brave sickly grin, the polite acceptance, even a tentative spoonful, and then the debacle!

We picked up the light off Jurien Bay, about half-way to Fremantle (having celebrated my birthday and Christmas Day en route), and I left Bob at the helm while I went below to sleep. And lying on my bunk I heard *Sheila* talking; it is then she passes her messages to me, some of which can be heard at no other time. Eventually at sea you get to know all the noises in the ship, and when you know their meaning they become unnoticeable except that they tell you all is well. For the first few days of a voyage there may be the clink of a loosely stowed tin in the supply locker, and it is traced and repacked into quietness; but any other discordant noise of the ship herself becomes a cry that will deny all sleep or peace of mind as effectively as a neglected child.

Now as she butted into each sea, I heard a soft hiss. So I rose from my snug blankets and searched for this noise, which led into the fo'c'sle. I pulled up the floorboards and along the upper seam of the garboard, near the mast-step, a solid jet of water hissed in from the sea each time the mast gave its downward thrust.

The leak was not as bad as the one on the previous voyage and we would have got to Fremantle safely, but this leak meant more to me than that—it confirmed the original uneasiness, it told me that in my desire to get home I was cutting this effort too fine, and that if I wanted peace of mind I must submit. So I went above to Bob, apologised for denying him his ride to Perth, took the helm and turned with the wind to Geraldton. That was a far bigger decision than a mere change in direction, it was the decision to stay in Geraldton another year because the southerlies were reaching their full force, and by the time they eased, the winter conditions in the Australian Bight would be too boisterous for *Sheila* and too hard for me.

I knew at once that my decision was right whatever it was to bring, but thought unhappily of the cable to be sent home, a home which had already waited so patiently for so long. I stayed at the helm all night and we covered the ninety miles back to Geraldton in fifteen hours.

After arrival in Geraldton I settled *Sheila* securely and delayed the expense of slipping as she did not leak at rest, and my first concern was to arrange a job within the crayfishing industry. There was big money to be made in this, but entry had not been possible the year before as a stranger. By early January I had arranged a job, but as the season did not begin until the middle of March I sought temporary work to keep me and increase my savings in the meantime. The wharf was still there, but that was a life I would not consider for reasons already given.

Apparently high wages were being paid far up in the north-west, mainly because the area itself held no other incentive for anyone to go there; and I wanted to know why. So eventually I was engaged through an unofficial agency in Geraldton as a builder's labourer at £24 a week and keep—an occupation I rightly assumed would bring no social prestige but great interest. My prospective boss was a contractor to a 400,000-acre sheep station, and needed labour to assist in laying water pipelines, building great concrete and iron water-tanks, shearing sheds, men's quarters, and a house. One of the station hands happened to arrive in Geraldton at that time and took me back with him.

We slept in the open or under a half-finished shearing shed; our meals were cooked by the boss over an open fire and the dining-room was a slanting sheet of tin, the dusty sand the floor. There was no fridge, nor even safe, so the meat was always blown and often tainted, because even cooked it lasted little time in that climate. Such vegetables as we got came from the homestead when a weekly truck called from Geraldton with stores. As fifty per cent of the vegetables received were rotten and slimy, the housewife threw these into the garbage tin outside the back door, from which the boss collected them to supplement our keep. It sounds terrible, but being ravenous and never stopping for morning or afternoon tea, I have seldom looked forward to meals more nor enjoyed them so much.

My job apart from carrying bags of cement and materials was mainly to tend the cement mixer, standing by it in the heat covered in sweat, dust and flies, pouring buckets of water into the revolving drum, shovelling in four of shells, fourteen of sand, five of cement, letting it mix, tipping the drum to fill a barrow, wheeling this to the construction site, returning, and the same all over again. We averaged a twelve-hour day, sometimes continuing by lamplight until ten or eleven o'clock at night, and to midnight once a week when we drove the heavy truck twenty-five miles for a load of shells for the concrete.

During the day I worked in a brief pair of shorts and tennis shoes, because clothing is harder to clean than the human skin, although at the end of each day most of the dirt stayed on us. The fine cement dust caked on to sweaty skins and into the hair, and even when the local water supply was turned on it was so brackish that any soap lost ninety per cent of its cleaning power.

Later we were moved on to building a new house for the station manager. I was fascinated to observe (and obey) the wide range of means by which a contractor can save the essential time, and yet with the outside boarding and inside wall-covering in place and smartly painted, it did look attractive. The station manager did of course often visit the site, but I was carefully primed on what action to take and what lies to tell for those occasions. It posed an interesting problem as to where loyalties lie.

The boss himself was completely tireless, and each morning we rose before light to his raucous yell, 'To horse, to horse, you riders of the purple sage.'

We rose to this on Sundays too because it was a seven day week.

One evening the boss said to me in friendly conversation, 'You know this place has been let go so much that there's £5,000 in the next few weeks if I could do the job.'

'I hope you do,' I said.

'I can't, Aid; I can't get more labour up here, not who'll work anyway.'

'I do admit I'm not surprised,' I said. 'No man of average intelligence will live under conditions like these when they are so much better in the south.'

'But look at the wages. Where will labour get wages like this in the south?'

'We're only working for the basic wage, Bill. We do an eighty-four-hour week, you knock off four hours for the keep, and that leaves £24 for eighty hours which is the same as £12 for forty, and not nearly so comfortable.'

'Gosh, I never thought of it like that,' said the boss.

Here another member of the team chipped in, 'Not bloody hell you didn't, you cunning bastard.'

That remark gave some cause for a temporary disturbance, and much to my relief I knew that exploitation was reaching saturation point and the team was breaking up. So did the boss.

It was not worth going all the way back to Geraldton and finding another job to tide me over until the crayfishing began, and I would be lucky to save £6 a week there whereas I was saving four times that amount labouring in the dust.

The heavy truck pulled out that night and I never saw it again.

Next day I moved my bedding roll over to a tin shack which I shared with some coloured people who also worked on the station. Most of these had a strong mixture of Malay blood from the Malays who had been on those coasts for years, mainly as pearl divers. The general conditions with those people were a great improvement on those I had vacated.

The work on the station was mainly making gates, fencing, cutting fence-lines through scrub, marking lambs, and carting water in a big truck. The average temperature recorded over the fridge in the homestead was 112 degrees, and heaven knows what it was in the open sun. The hottest day recorded was 123 degrees, when I was out clearing a line through the scrub. This job consisted of hacking down a bush and then lying exhausted for half an hour in the shade of the next. The boss came out in a jeep after lunch, took me home and gave me a bottle of beer straight out of the fridge. He was the only employer I had in Australia, apart from Mick, whom I'd willingly work for again.

That station was a strange place, and for looking after the huge concern the manager's wage was only £25 a week while the owners lived in Sydney. It ran 13,000 sheep, but how the first white man looking over that country had the audacity to believe that sheep could live there is hard to understand. Sand, thorn-bush, and very little rainfall. Water can be had from artesian bores in many places, but the sixteen once erected on that place had fallen into disrepair and only one remained at the homestead. The impression gained was that in times when the price of wool was low the loss of a couple of hundred sheep wasn't worth worrying about; and when the price of wool was high, well, who with 13,000 sheep worries then about 200?

Long pipelines fanned outwards from the central bore to distant troughs, making long curved arcs in the sand as they expanded and contracted with changing temperatures. There were few fences, dogs could not be used for mustering because the sand was too hot for their feet, and much of the scrub was impenetrable to a man on horseback. The sheep were mustered simply by turning off the water that led to the troughs, so when the animals called at their particular water-point to drink and found no water, they waited near the trough in the hope that it would come. I went out to round up one mob after the water had been turned off for about a week, and among the thirsty sheep were equally thirsty emus and kangaroos.

The flocks were then driven into the homestead paddocks and yards for shearing and marking the lambs. The water supply was not plentiful there either, and each evening another man and myself dragged out the bodies of sheep which had died of thirst, loaded them on a truck, and drove into the scrub to burn them. Sometimes we missed a body for a couple of days, and when we eventually grabbed a leg it tore away from the rotten carcase and made further handling unpleasant. I hated that job.

Several times I went into the small town of Shark Bay, which can best be described by saying that not a well-kept plot of land nor a garden fence was to be seen. The houses were tin or weather-boarding, and although fortunes were taken from the fish-infested sea which lapped the main street above the beach, no evidence of wealth was seen. I was told that few, if any, locals fished consistently, and when a catch was made the crew lived ashore until they ran out of cash at the pub. There were always plenty more fish. I don't think the Australians will ever settle that part to any extent, and any race who does will still find the south more attractive. Thousands of New Australians, who generally work hard and drink less, are already thriving in the south.

Late in February after two months in that area I drove my leg against the broken branch of a tree reputed to be poisonous. I seldom go bad, but within two days the leg became swollen and the glands in my groin very painful. Fortunately for me at this time a local domestic crisis came to a head (women, any women, were gems of rarity in that country), and as far as I could see everyone just walked off the place. The shearers loaded up their covered lorry and offered me a lift to Geraldton, and after calling at Shark Bay for a cargo of refreshments we headed south. An empty bottle flew over the stern every half-mile, blazing the trail for others who might be so unwise as to follow it back to the north.

We drove into Northampton in the early hours of the next morning, and I slept on a bench outside the deserted pub as the whole town was asleep. Later that day I arrived in Geraldton with £150 saved, and admitted myself straight into hospital as by that time I could hardly bear any weight on my leg.

The next week was bliss—complete rest, clean sheets, the Matron was efficient and the nurses pretty; and although the early awakening imposed apparently in all hospitals prevailed, it was more gentle than that raucous voice, 'To horse, to horse, you riders—.' On the seventh day my new skipper called for me in a truck, and that night I was on a launch headed for the islands.

34

Crayfishing—wrecked—out of a job

The crayfishing industry of Western Australia extends along the 400 miles of inhospitable coast from Fremantle northwards above Geraldton, and is now Australia's biggest dollar-earner after wool. This marine wealth was hardly touched before the War, but the higher post-war wages paid in the USA have brought this luxury within the purchasing power of the average American working man and his family. It is this huge market which gave the industry the stimulus it needed.

My job was with a carrier vessel running from Geraldton to the northernmost of the Abrolhos Islands, the scruffy top of which I had seen in the face of the sinking sun that calm Sunday evening almost a year before. The vessel was owned by one of the processing factories and was leased each season to a skipper who made his own arrangements for a crew, in this case me. The charge for transporting crays on that eighty-mile run was roughly £1 a bag, of which the factory took half and the remaining half was shared equally between the skipper and myself. If we carted 100 bags on a trip the factory took £50 of the transport costs paid by the fishermen, and the skipper and I each took £25. At the height of the season we expected to do three or even four trips a week. The skipper was pleasant and easy-going and had given me generous terms, but after two trips I spoke of two points which needed attention. The first was that we carried no stern light, a serious defect because when laden we had no view aft from the wheel. The two other carrier vessels on the same run as well as coastal shipping were all faster than us, and the thought of another ship climbing over our stern one dark night disturbed me.

The second point was our compass, for which there was no deviation card (courses taken from a chart are true in relation to True North; a deviation card shows what allowances must be made to ensure that courses sailed by the ship's compass correspond to the true courses). The type and position of the compass was also such that accurate bearings could not be taken from it to outside objects. It was about chest-height under a small naked light, and at night with spray on the glass of the wheel-house all you could see ahead was the reflection of the compass and its light. This not only hid any view through the glass, but the compass light itself blinded the eyes to normal night vision.

I talked to the skipper about these points but he was not over-worried.

'She's right,' he said, and I'd learnt that those words uttered in the Geraldton area meant that further pursuit of the matter was a waste of time. These were simply needless risks I had to accept in return for the good money made, but other factors began to arise which made me consider leaving the job of my own free will.

As catches increased and our work with them we often ran in from the islands one night, unloaded, reloaded with stores, and left again the next night; and that vessel was not like *Sheila* to steer, she was a bucking bitch of a thing that had to be held on her course. It became obvious that the lack of sleep and erratic meals—that is, fatigue—was going to be our greatest hazard, and it became imperative that such time as we had ashore should be organised to recover what we lost at sea. But the organisation was nil.

I didn't mind the long hours and the hard work but I did resent the stupidity of increasing fatigue unnecessarily, because this meant inefficiency and the loss of all joy in work, and that at sea sooner or later means disaster. I didn't want this job to end because it was my means of getting home.

This 'she's right' doctrine which the local Australian mistook for his cherished individuality and clung to as his most valuable national characteristic was also his greatest enemy; because each individual refused to be co-ordinated, nothing was co-ordinated, and the result was the appalling administration with its inevitable cost of time lost, energy wasted, and often personal friction. There was seldom retribution beyond this because of their country's enormous and easily acquired natural wealth, and their own undoubted self-reliance and ability to improvise. But in the comfort of that simple expression lies a far greater threat to Australia's future than that to be expected from the yellow youth growing to lusty manhood in the north.

Over the next weeks the work increased and we became weary. One night when we happened to be carrying a fisherman passenger, I handed over the wheel to the skipper and went to my bunk. Soon the passenger awakened me, 'Hey, Aid, wake up. The skipper's crook.'

In the dim light of the compass I saw a huddled form on the floor beneath the wheel. I brought the ship on to her course and shook the form into consciousness.

'What's the matter?' he asked.

'Nothing's the matter with me,' I said. 'What's the matter with you? Are you crook?'

'No, I'm OK. Just felt a bit sleepy—she'll hold her course for quarter of an hour, when I wake up and check.'

I knew these quick naps very well and that they are very liable to continue into a deep sleep lasting for several hours, so that night and others I took the wheel in self-defence until approaching Geraldton. That week I wrote to a friend in my regiment describing this way of life much as I have done here.

'At last I'm making good money,' I wrote, 'but if we keep going like this it can't last, because we're going to hit something very hard, yet there seems nothing I can do about it. My biggest worry is that I'm going *magra* (a Gurkhali word meaning something between sullen resentment and frustrated fury).' I didn't know then that the problem was to solve itself within the week.

One night we loaded to maximum capacity, 160 bags (about six tons). They filled the cockpit, covered the engine casing, were stacked in the cabin and on our bunks, and on the foredeck. Luckily it was a calm night, but orders came over the two-way radio to off-load sixty bags on to another carrier en route. This took us off our normal course, and because we did not know the deviation of our compass (we could have made out a card in an hour during any of our visits to Geraldton), the true courses taken from the chart were a poor guide.

That night we hit three reefs before we finally cleared the last island by midnight on the open course to Geraldton. By that time, having poked our way among the reefs for hours with me sitting on the bow with a spotlight, we were both feeling the strain.

I took the wheel till 2 a.m., called the skipper who woke me again at three, and I continued until we were about half an hour off Geraldton.

'The leading lights will come up in about half an hour,' I said, and without waiting for his opinion went to sleep on the engine casing beside where he stood at the wheel.

I awoke to a crash and a speedy take-off from the casing. The ship picked up again, came to another shocking check, rose on the surge and shot ahead a few yards, and then the skipper throttled back and we came to a halt on the coral, listing heavily, and each surge swept over the side-deck into the cockpit. It was still pitch dark but I knew we couldn't sink, which was a comfort. The lighthouse was immediately above so I flashed an SOS, and hoped we'd be rescued soon because we were soaked and bitterly cold.

Launches came out soon after dawn and we were finally dragged right over the rest of the reef and towed into harbour. The whole cargo of crays, over £1000-worth, was condemned at the factory because the bags had been contaminated with diesel oil—a loss of about £25 to me because we were not paid for rejects.

As far as I know there was no formal enquiry, even by the factory which owned the boat; anyway my opinion was never asked much to my relief. Everyone was very kind.

'Hard luck, chum, it could have happened to anyone,' and under that kindly outlook the hard glaring causes of such an inevitable effect were lost. But the ship was badly damaged and I was out of a job, in the middle of the season when another would be hard to get. It is strange how sometimes our wishes are fulfilled, but seldom in the form we expect.

35

*The Trimmerwheel—a girl—an offer for Sheila—
departure from Geraldton*

A few days later as I wandered about the wharves, a tall good-looking person of about thirty approached and asked me if I wanted a job. He was skipper of his own boat, the *Trimmerwheel*, a big sixty-foot twin-diesel launch which I had often seen running in competition to my previous ship. I simply hate either asking or asserting what the value of my hire might be, and so it was not until we were on the way to the islands that I raised this embarrassing subject.

We finally agreed on a fixed wage of £18 a week plus my keep while on board, and I was also to live on board while in port. This saved me £5 a week board and lodging ashore, and suited Fred because he lived some way out of Geraldton—it often blew up suddenly during the night when a boat lying alongside the wharf had to be moved out to a mooring without delay.

'We'll share the work,' he said, 'but I want the ship kept clean and tidy inside. I'll always get the big evening meal at sea, but one thing I won't share is the washing up. I hate it.'

'So do I,' I said, 'but that's fair enough.'

Living on board the *Trimmerwheel* was rather like occupying a self-contained luxury flat in London. She had a big panelled saloon with comfy settees, a table, lockers and book-shelves; a neat, well-equipped galley, more bunks in a cabin for'ard, and a shower, hand basin, and lavatory; running water and electricity.

Over one period we had a lot of engine trouble, which kept us in port for several days. We worked almost every night until ten o'clock or later, which I mused was not according to union rules, but Fred made this up in many ways, such as by allowing me the use of his car, and off-parade we found we had much in common. This is perhaps the ideal relationship between employer and employee, but it balances on a knife-edge and can be easily destroyed by either. Just as an employer can spoil labour by being too generous so can an employee spoil his boss, to their mutual discomfort because giving too much is as dangerous as giving too little, and the middle way is hard to find.

There were rough trips, and nights of plugging into the hard southerlies when the vision through the spray-covered glass of the wheelhouse was nil, but there were days too of glorious sunshine when we steered from the flying bridge, and sometimes left our course to chase the whales passing through the channel at that season. Often a huge body launched itself high out of the sea to land with a splash which could be seen ten miles away. This was their courtship, but further intimacies were discreetly hidden beneath the waves.

One night of a half-moon and squally winds I was steering from the flying bridge, and saw an area of white water straight ahead. We were far beyond the reefs at that time, and thinking the disturbance was a mere scuffle of wind I held the course and drove straight through it. There was a violent bump, a loud swoosh of rushing water, and the long tapering tail of a whale rose beside the ship, and the wide flukes almost touching the short mast were etched against the moon-soaked clouds above my head. It seemed to hover as we bumped across its back, and then plunged headlong out of sight sending clouds of spray over the ship.

Fred had business in Perth, but with his carting contract to fulfil he asked me to take the *Trimmerwheel* over to the islands. I had learnt something about the diesels during our days of pulling-to-bits and putting-together, but it had not made me love them any more.

'How do you feel about the engines?' he asked.

'I'll press the starter button when I leave here,' I said, 'and shut them off when we get to the other end; in between I won't look at the bastards.' This remark evidently instilled little confidence in my mechanical skill so he sent his uncle with me. He was an engineer, a good cook, a keen fisherman, and a most pleasant companion.

We anchored at North Island that night and fished all the next day, loaded up the following afternoon and left North Island just before dusk. The low flat line of Pigeon Island showed on the horizon just before it became too dark to see it, allowing me to check the compass course; it needed no check but after darkness falls it is always a comfort to have done so. When the lights on the fishing launches showed up inside the lagoon we entered the reefs on bearings, and took more crays on board while Earle cooked up a big steak supper.

The plan for getting out was by that time almost routine, and with confidence in the details concerned it held little risk, but passing through black night-covered waters, visualising the chart and knowing the reefs sliding past invisible on either side, never failed to bring a thrill. These are the times when you most feel one with a boat, as if the two of you are entering a naughty conspiracy with a mutual confidence in getting away with it.

Earle cast off the mooring, I turned *Trimmerwheel* and headed down the dark lagoon with both engines murmuring at 600 revs, and picked up the last pontoon in the searchlight. This was shut off as the pontoon drew abeam and only the shaded compass light remained. I looked at my watch and swung on to a course of 81 degrees for three minutes, altered to 102 degrees for 2½ minutes, and then in a wide easy turn to 165 degrees.

A young fisherman by my side said, 'I suppose I'll always be working for some other bastard, because I wouldn't be in your shoes now for £1,000.'

'It's not my boat,' I said.

After five minutes I knew that we had passed through the fringing reefs and I pushed both throttles forward to give 1300 revs, holding that course for Geraldton and leaving North-east Reef clear on the port beam.

Soon after one o'clock in the morning we picked up the loom of Geraldton lighthouse, and as we rose over the bulge of the horizon it winked out its steady rhythm before the bow, and two hours later we turned down the path of the leading lights, picked up the winking channel buoys, turned again and reduced speed as we passed through the breakwater into the harbour. Then in moments we were snug alongside, I shut off the engines and uncrossed my fingers, had a mug of hot tea and crept gratefully into my bunk. In many ways handling someone else's ship is far more worrying than handling your own, but I must admit that engines make it much simpler.

At this time I went mad and spent £100 on new clothes, and it was the most sensible thing I did in Australia. Bad clothes mean bad morale, which is only another name for unhappiness.

One evening I heard a car draw up to the wharf against which the *Trimmerwheel* was berthed, and a man called out,

'May we come on board?' And down they came—a recent friend and two girls. I got out a bottle of Rhinegold, a pleasant Australian wine, and it was heaven sitting there in a decent pair of slacks and a fresh shirt, talking; these were almost the first girls I had talked to since my arrival. After the months of living in tin shacks, the heat and cement dust of the north-west, the lack of finesse in the preceding jobs, gutting fish and sleeping in stinking clothes, to see two girls carefully dressed, so naturally clean and discriminate, made me remember so many things I had been in danger of forgetting. Sometimes a whiff of perfume made my senses reel, and the wine was kind.

The elder girl was the more thoughtful and sophisticated. She had travelled a lot but it seemed that her easy self-assurance was in part a refuge, because behind the laughter there was sadness in her eyes. The younger girl as yet had no refuge, not having known the cruelty that makes one necessary. Behind her chatter and laughter were fleeting glimpses of wonder, as if all she saw and heard carried a subtle wondrous message which she could not quite grasp and hold.

It was not so much a direct attraction I felt for her but more as if a great magnet lay equidistant between us, and only by meeting in that would the pain of attraction be stilled.

But at one time when I was much younger and had loved another deeply, I had thought that neither of a pair could feel this powerful attraction without the other also feeling it; and as we lay together in complete contentment I had said, 'What is this heaven, that when we touch all time is lost, and nothing is known but this deep stillness?'

And she had propped herself up on her elbow, and looking at me with a fond, amused smile had said, 'What on earth are you talking about?' And I had known it no more for many years after that; so how could this young and lovely girl to whom I'd barely spoken be aware of it?

Soon she had to go and my friend said,

'You take her home, Aid, the keys are in the car.'

And driving back in the car there was a great awkwardness between us, because my casual conversation was blurted out against that which possessed me, and which could not be spoken because to utter it for rejection is to commit its desecration.

I stopped the car outside her gate, and for some reason neither of us moved. The pale light of a distant street lamp filtered through the dusty windscreen, showing her face strangely clear in the dimness inside the car. I heard myself saying, as if with a great fear, 'Is it allowed to kiss you good night?' And the silence, and her faint 'Yes' as she turned towards me.

I bent and kissed her lips, and after so long, after so much, knew again the firm softness which defines a woman's lips beneath your own. She lifted her hand to my face, an impulse of sudden tenderness, to give her lips a little longer. Then swiftly she was gone.

To kiss a pretty girl good night is nothing uncommon, but as I drove back to the boat I knew that this was a turning point in my voyage. The bitterness I had felt in Australia, the exhaustion, the worry, the doubts and regrets and loneliness, all fell from me with that simple gesture given and received in trust.

In this I am sure lies man's greatest need of woman, but in seeking the more obvious attraction of the opposite sex we take the lesser and so lose our perception of, and our vulnerability to receive, the greater. And by becoming unreceptive we men forbid women to express it, which is their great need as it is equally ours to receive, and in its manifestation only does each become both; without expressing it women die, and without receiving it men become callous, offending themselves and all the world.

I signed off in the latter half of August, having made £500 since the preceding March, and began preparations on *Sheila*. I took time less seriously and seldom hesitated to go sailing with friends in the glorious weather that prevailed. Those were happy days; my whole attitude changed and very nearly ended by my staying in Geraldton for good.

There were many friends, not only in the well-dressed Saturday evenings but during daytime work. By that time I knew many of the fishermen, some of the lumpers, Jeff, and odd-job men around the wharf, and under their casual, superficial air I found that all these were generous and good-hearted people it was going to be hard to leave. I began to understand that way of life, and one day I found myself walking barefoot through the town only in a pair of those brief shorts Australians wear, which on arrival had made me wonder how anyone could bear to be so undignified! I even heard myself using that expression 'She's right', and by accepting its dangers I found its comfort also.

As *Sheila* lay alongside the wharf one day, a well-dressed, middle-aged person stood for a long time looking at her, until he asked me,

'Are you Hayter?' I confirmed that and he went on,

'I want to buy her. How much do you want?'

'I'm sorry,' I said, 'but she's not for sale.'

He then explained that he had just retired and wanted a boat. He could buy any boat, but he only wanted *Sheila*.

'You can ask your own price,' he continued, 'and I'll take her over as she is, immediately.'

'No,' I said. 'I'm sailing her to New Zealand, as I've always planned to do. Anyway the import and sales tax here is so heavy to get what I wanted I'd have to ask something absurd, like £5,000.' And quick as a flash he asked,

'Will you take £5,000?'

All the reasons for wanting to stay in Geraldton passed through my mind. With £5,000, after paying the taxes, I could set myself up with a good one-man cray boat, gear and equipment, and a second-hand car. Anyway I could bump him up to £6,000. After three years' steady work I would have a better boat, a better car, my own home, and be married.

'No,' I said again, 'it's not worth my while.'

'Will you tell me how you work that out?' he asked.

And because I didn't know what sort of a person he was, I grinned to make it polite and said, 'No, I won't.' So he gave me his card, saying that if I did change my mind to contact him immediately and that I promised to do.

That restful and happy period in Geraldton I recognised as the break before the third and last round, when the greatest effort is demanded and temptations come crowding in to deny the success almost within your grasp. My physical arrival in New Zealand in *Sheila* would matter nothing in itself, but the failure to do so would, and no sum on this earth can compensate for that. If I failed through hitting a reef, or being run down, or through inefficiency or just giving up, or accepting £5,000, what's the difference? It is all failure.

And so he went away and I got on with my work, and eventually sailed for Perth in early October 1955. I entered Fremantle four days later, and so fulfilled the belief that had come to me nearly two years before.

36

Arrive Fremantle—the Yacht Club—more crayfishing—an unlucky ship

On arrival in Fremantle I took *Sheila* up the Swan River to a sheltered and clean berth at the Royal Freshwater Bay Yacht Club (RFBYC), who had kindly invited me to do so. It is unpleasant living on board in a busy harbour because everything becomes filthy with dust and grime, the wash of passing tugs batters *Sheila* against the wharf, and living within the environs of an industrial port for me always brings an atmosphere of human depravity. I wondered greatly whether this was no coincidence; the edge of the land feels most strongly the contrasting cleanliness of the open sea, and perhaps unconsciously summons its evil to resist the submission that cleanliness demands.

The RFBYC is the most beautiful and best equipped that I have seen on all my travels. The large clubhouse stands in wide lawns crowning a small spit of land, around the base of which are jetties, slipways, workshops, stores, and dinghy shelters; breakwaters enclose the pens where larger and varied craft lie in safety. A few evenings later I was asked to attend an official welcome by the club, and give a brief talk on my voyage; and I gave one of the worst I have ever given.

These men were all keen yachtsmen, and the racing in which they delighted had undoubtedly given many of them far more finesse in sailing than my cruising had given me; but from their ocean racing they knew quite enough of the sea to appreciate the physical and mental demands of a single-handed ocean voyage. These were the problems in which they were interested but which to me are merely a means to a greater end in which my real interest lies. I tried to force my interest into these expected channels but I felt I was being a sham and insincere, and if a speaker is insincere he will never establish that true indefinable contact with his audience, however interesting his facts.

I had arranged for a job on a launch for the short white cray season, the preparatory work for which began in mid-October, but that was then two weeks away and the interval I filled by working in a wool-store in Fremantle at three guineas a day. Each morning I cooked breakfast and packed my lunch, walked through the impressive portals of the RF-BYC with a wool-hook stuck in my belt, caught the Fremantle bus and reported with a motley throng for work. And each evening, my clothes greasy and reeking of sheep's wool, I boarded a bus full of neatly dressed typists and shoppers, and returned to my floating home. This was the only fixed wage job I had in Australia where I earned every penny of it.

I then moved out to live with the family of my new skipper to help prepare the pots, floats, and gear for the crayfishing. The third member of our crew was Ted, a newcomer to the area who knew nothing about crayfishing and who had accepted a rate of £10 for every 1000 lbs. of crays taken (I had accepted £12). On these low wages the skipper was very lucky to finish the season, and indeed caused himself a greater loss than if he had paid higher… a statement that can never be proved. But these stories get around and such a ship will never have a happy crew; and a ship without that is a ship without luck.

The white crays are identical to the common red variety except that they are pale pink in colour, but otherwise little is known of them. They are barely seen outside that one six weeks' season in the year, when all along the 400 miles of coast from below Fremantle to far north of Geraldton they appear overnight on the same night. They are said to come in on the surface from far out at sea, and they settle in about five fathoms near the coast and then they travel back along the bottom to deep water, moving out about half a mile a day until they disappear and are not seen again until the next season.

Many stretches of that long coastline are untouched except by a few big boats which contain their own freezing plant, because the fishing has to take place within reach of the isolated processing factories. We sailed in mid-November for one of these situated about 100 miles up the coast, and dropped anchor inside the reefs behind a small island.

Thereafter the routine was always the same, Sundays included, except when a merciful storm made conditions too rough to get out of the reef. We rose at 3 a.m., had such a breakfast as is barely possible at that hour without retching, and steamed out through the reef just before dawn guided by two lights stuck on poles in the sand-hills; and having hauled and reset our 100 pots we turned for home about 3 p.m. with the inevitable hard southerly under our tail. Ted and I cleaned down the ship, had a cup of tea and huge chunks of bread, changed into dry clothes, and relieved the skipper at the wheel.

On arrival inside the reefs we unloaded our catch, pushed it on a trolley to the factory (a forlorn tin building in the sand-hills), checked the weigh-in and the docket, collected our stores, and returned to our mooring.

For the rest of daylight we sometimes dived overboard to swim in the warm, clear water, or there were pots to repair, rope or floats to splice as each day the crays moved out into deeper water, and while the skipper got the evening meal we often worked till after dark. Ted and I took turns at washing up but we were all usually in bed by 9 p.m., dead tired after an eighteen-hour day, every day of the week. It seemed only seconds before the alarm went and it began all over again.

Sometimes there were breaks in the routine. Each pot was in itself an expectancy, and sometimes a wave lifted the spinning screw over a float and one of us would put on goggles, take off his clothes and go over the side with a knife to clear the rope. It was a strange, lonely feeling far under the hull of this big craft, rising and falling in the seas, the spare tail of rope snaking in the suctions. We were often twelve miles from land in about ten fathoms, and through the clear sunlit water the sandy bottom and the sinister reefs were clearly seen far below. I was always very conscious of my legs dangling below the hull as I worked frantically to clear the rope before my breath ran out and once, five minutes after climbing back on board, a hammer-head shark cruised alongside.

For some reason our catches were not nearly as good as others', which aggravated the situation. Tempers were becoming short, and I have come to believe that in ninety per cent of cases this is usually something to do with the stomach. We had nothing to eat from 3 a.m. until we had a cup of tea twelve hours later on our way home; any man can go without food for this time when it is necessary, but only a fool does so when it is not. No fire can burn brightly without adequate fuel.

Ted and I both got touches of cray poisoning, when cuts from the cray claws and shells fester and throb painfully at night to deny sleep. This only added to the trouble—all the signs of a break-up were so apparent, so unnecessary, and the solution at that stage so simple.

There was a row on board one night which rapidly approached a point beyond verbosity, and a fight under those conditions can end anywhere. Before leaving next morning Ted said to me, 'Now do you see why we have our unions and make these bastards pay?' (He belonged to the powerful Seamen's Union, and some of the tales he told of what they got away with amazed me.) 'There's no friends in business, Aid, I can tell you.'

That of course is a common expression and I had always thought it rather cynical, but only later saw it to be true. Business and friendship are at one and the same time incompatible. If someone sells me something at a large profit to himself, that's good business but is certainly not friendship; if he sells me something at a just profit, that is business but neither antipathy nor friendship although it is that out of which friendship may grow. And if he sells me something at a loss to himself, that may be friendship but it certainly is not business.

After Ted left I took on his work as well as my own, with his wages. Soon after Christmas, near the end of the season, the finale came. It was always the skipper's job to get the evening meal, which he did exceedingly well, but when I came down at dusk from cleaning up on deck I saw he had made no move and we were both famished.

'What about supper?' I asked.

'I haven't a clue,' he said, and so I went into the galley to get my own.

'I'm running this ship,' he yelled, 'and you'll bloody well wait until I'm ready. You're doing very lightly on your wages as it is.'

'The sooner you do run your ship,' I yelled back, 'the sooner you'll get people to work for you. We'll make up my account tonight and I'll get to hell out of it first thing in the morning.'

And so on that childish note the season ended, not because the work was too hard nor the hours long, but simply I'm sure because of poor administration which momentarily changed even our characters. I cleared £170 in the ten weeks and the total value of the season's catch was about £1100, but after paying running expenses I doubt if the skipper cleared much more than £500. That's not bad for ten weeks' work, but if the catches made by the other boats were anything to go on, it should have been double that. Our boat was just unlucky.

37

Perth—the payment of bills—the clean handkerchief

On return to Perth, a few miles up the Swan River from Fremantle, I immediately concentrated on the final preparations for departure, which as much as anything consisted of picking up the threads of arrangements initiated before going crayfishing.

One of my greatest difficulties in finalising departure has always been to collect bills owing; until I do know what these are I cannot know how much cash I can spend for the last item of supplies. In Perth, in desperation to save time, I ordered the barest minimum required, painted the tins against rust, stowed them and paid the bill, hoping that I would have enough cash left to meet commitments still outstanding, but for which I could not get the bills. This unknown future always made the last days in port a hell of anxiety.

After many experiences in many ports I had come to know that these bill-withholding tactics sprang inevitably from one of two motives. The first was because the bill to be presented was exorbitant, and by presenting it at the last moment the merchant concerned knew that rather than delay my sailing to argue the point, I would pay it in full without demur. These wily creatures somehow seemed to know that I was not the type to slip away without paying it; or if I did, they would have no hesitation in claiming the sum from the Yacht Club which in their commercial eyes had sponsored me.

'I trusted this man,' they would say, 'because you made him an honorary member of your club,' and the club would pay to shield their name, and I would never hear of it.

The other motive was exactly the opposite, by a person who wanted to make a gift and such people do not want it known. These gifts were sometimes made by a person who had all his life wanted to do what I was doing, but never having quite got round to it he then made his contribution to me. Others perhaps knew I was short of cash and in generosity bore the cost instead of me; and these were often people who in meeting obligations which I had sacrificed were as hard up as myself. And the last were those who were consciously aware of what I sought, and in making a gift to the lesser me they somehow made a contribution to the greater That. This to me is the greatest gift of all because it carried me with it.

In this last category were also those who came almost secretly to *Sheila* and asked me to accept some small thing 'for good luck', or being afraid that might be misunderstood, some quite useless thing that they hoped might be useful.

I valued all these things, because to wish another good luck sincerely is really a prayer and such prayers are answered. It generates this power greater than the giver or the receiver, placing them both in some unfathomable way within its protection. Many of the people who gave me those small useless things knew I needed money, and some I discovered in various places were millionaires; but they also knew that money carries its own reward, and so has not the power of the useless things.

I loved those last days in Perth, because perhaps only then did I fully realise what wonderful people there are in that city. Official occasions in the club were conducted with the ceremonial of formal occasions, because ceremony is the hard superficial shell which protects the vulnerable kernel of any institution, but beyond that formality their true natures came to the fore and staggered me by their kindness and generosity.

For instance one of them (perhaps initially because it was the thing to do) asked me to tea at his home (dinner in England). We sat on an open verandah with a cold beer, had an informal meal, and sat again afterwards talking until quite late. And when I left they said,

'Now come up any time. We don't need any notice because one extra meal is no trouble at all. If we're out, here's where we keep the back door key; just walk in and look at the papers, write, listen to the radio, what you will. There's always food and a cold bottle of beer in the fridge.' And once as we talked, hardly aware of the children playing beside us, I was saying that I always remembered to keep a clean pair of slacks, a shirt, and socks, in which to go ashore, but I always forgot to keep a clean handkerchief so the effect was often spoilt when I pulled out some filthy bit of rag. What an effect that statement was to have—the eldest girl of about seven paused to regard me steadily for a second, and then continued with her play as if it were already forgotten.

The Yacht Club shipwright did a lot of work on *Sheila* in a fifth of the time it would have taken in Geraldton. He kept me waiting for the bill until the last day, and with my suspicious mind my last request for it was barely under control.

'Forget it, Adrian,' he said, 'I'm glad to help.'

Sheila was slipped for a last anti-fouling, a hard, long job by myself but one to be accepted with the joy of owning a ship alone. Yet members of the Club turned up with no request from me and the job was done in a matter of hours only—in this 'You're in Australia now—' country, where everyone helped himself!

People gave me warm clothes, cakes to take with me, and the dentist even pulled two teeth as his contribution.

Then the last bill wrecked me. I had asked for it four times over the preceding ten days, and the last morning the manager himself brought it down. The pattern followed along alarming lines, when the behaviour and the words are identical whether in English, Hindustani, Portuguese or Australian—there are no basic differences between races but a world identity of types. After the expression of his personal admiration of me the manager stated the eagerness of his firm to help yachtsmen, so I knew my careful estimate of about £7 was to be incorrect. The bill was £30; I paid it without a murmur, and didn't bother to load *Sheila*'s gear from where it lay beside her on the jetty. I had about half a crown left.

I could go back to the wool store for a week, but delay now seemed to carry the foreboding of a turning back. I wracked my brains for something to sell, having already sold my cine camera which the Customs had been entirely wrong in putting on their 'frozen' list, but everything else of value was. And suddenly it came in a brilliant flash—the ship's lavatory; it was not on the Customs' list and it was the best brand, costing about £80 new.

I'd never sold a lavatory before and was somewhat embarrassed by my entry into commercial life with such a personal article. However it brought £30 and left me marvelling greatly at the scheme of things, whereby a venture such as this ends with the aid of a bathroom appliance.

In my list of goodbyes was the little girl who had overheard the conversation about my going ashore difficulties. She brought me a freshly new handkerchief, in one corner of which she had herself embroidered a large 'A' in coloured silk thread.

'Now, Uncle Adrian, this is a clean hanky for going ashore. You must put it away carefully, and promise not to use it until you land in New Zealand.'

And I smiled at the very serious look in her steady gaze; children regard these small things as if they are a matter of life and death.

'Do you promise?' she asked searchingly.

'Yes,' I said, 'I promise.'

And she gave me the handkerchief, smiling with a quiet content, because in her trust her wish was already a fact.

38

*Departure Fremantle—Planning—the Roaring Forties—
a ship passed—Bass Strait—Refuge Cove*

The masts were re-stepped in Fremantle harbour. I finalised my clearance papers and completed the setting-up of *Sheila*'s rigging by dusk, when I went ashore for a meal. It brought a feeling of unreality, as if I were a ghost wandering through the town that night watching people about their normal ways, knowing that tomorrow I'd leave it all and head out into another world alone.

I was abrupt and barely civil to the reporters who saw me off. This was not deliberate incivility; it was partly preoccupation with the details of casting-off, but more my awareness of all that I was going into and all I hoped to find by doing so, and their queries and photographs, even their sincere good wishes for a safe passage, seemed an intrusion into some greater privacy even beyond my own.

Early on the morning of 10 March, 1956, after a hasty visit to the dentist to extract two more teeth which had suddenly begun to ache madly during the night, I headed out of Fremantle on the last leg of my long, long voyage. I felt jittery and on edge, knowing that this would pass in a few days, and also that I would find no love of the sea until it did.

My plan was to head south down the coast, giving Cape Leeuwin (the south-west corner of Australia) a wide berth to clear its strong on-shore currents, the danger of which the Pilot Book stressed in several places. In the thick weather which prevailed I never even saw the Cape. From there, to avoid the unfavourable winds which blew on the shortest direct route across the Bight to Bass Strait, I continued 300 miles further south to latitude 40 in the Great Southern Ocean, to pick up the westerlies that blow hard and consistently the year around. These are the 'Roaring Forties' so well known in the days of the old sailing ships, which now have gone, but the Forties remain for such as me.

In the first two weeks out of Fremantle I covered 600 miles to the south through variable winds and confused seas, to arrive in the vastness of the Southern Ocean. Each sea and ocean has an individuality of its own, as distinct as that of each country, and the impression here was not of physical vastness so much as the vastness of eternity, of timelessness. Sometimes the huge rollers, bigger than any seen before, were tipped in white and the hurrying wind nicked their crests slithering along *Sheila*'s decks; sometimes they were great friendly giants marching without haste, without pause, on their way around the world. At other times they broke their long lines and the vast sea became a flexible canopy lumping up in huge mounds, acres in extent, as if the giants had gone below to heave their shoulders and flex their muscles.

It was on my way south that my diary records one morning unique in all others, perhaps because of the many preceding years spent in the tropics. Its awareness came with a single soft puff of wind on my face:

> There's a feel in the air of winter, not of biting cold, of dampness and cloudy days, but a feeling of freshness like the promise of something wonderful, like spring but in reverse.

I had crossed latitude 23 (the official border of the tropics) long before on my way from Java, but this is where I entered the cool zone in which my own home lay. Leaving the tropics had been tinged with regret, but the wonder in that cool puff of wind banished that regret with the joy I found entering into my own. It is very strange that in this simple experience lay the secret to the great answer I sought—where from a first joy springs a regret at its loss, and from that regret is born the awareness by which a second joy is known, which annuls the regret and unites you with the first. But words cannot bear the burden of that meaning, only the understanding which lies secretly in us all.

It was over this great ocean and the winds that blow across it that I found the albatross, the solitary bird with scimitar wings, which in harmony with solitude breathes an impression of immunity, a mystical union with wind and sea. They sped for miles in the air-stream rising before each onward rolling wave, curved high to the sky with no movement, hung motionless at the height of the turn, and with a huge graceful swoop of slow speed came back to the sea to find with beautiful precision the rising stream of air before another roller.

In the third week out I entered into the full path of the Forties, put in three reefs and the storm jib and flew before the heavy winds at eighty to 100 miles a day. I averaged about eighteen hours of each twenty-four at the helm, and when inclined to do more it was not fatigue but cold which drove me below. Much to my surprise in those cold waters I saw many large sharks, which lay just below the surface, presumably to avoid the disturbance of breaking crests, and they only moved with a great swirl when *Sheila* ranged alongside. Porpoises were seldom seen but whales were fairly common, surfing down the forward slopes of the great seas with their huge, blunt heads half out of water.

Over other days and nights the wind rose to near gale force, flinging spray and broken crests on board, soaking everything below, until I had to creep between my sodden blankets still wearing my oilskins. This robs sleep of the refreshing properties found in dry sheets and silk pyjamas. In this time also I remembered without charity the very smooth gentleman in Fremantle, whom a drop of salt water would have sent to bed for a week, and who had sold me an expensive set of oilskins with dignified but penetrating assertion of their true worth. They were indeed waterproof, but the stitching rotted within three weeks at sea so that each arm and leg became an apron, and the parted seam down the back welcomed each boarding sea with open arms. The bastard.

The last day of March is entered:

> Started off at 4 in the morning and sailed all day until midnight—20 hours at the helm in real Roaring Forties stuff, finally under storm jib only. Hugh beautiful seas, few breaking and then

not badly, thank heaven. I could have sailed on till the next dawn but the cold was cruel.

And that day, three weeks out, I struck 1200 miles off the chart, leaving only 700 to Refuge Cove.

In planning this voyage I decided to bypass Melbourne although it was only a few miles off my route, partly because of the dangerous tides in the entrance to Port Phillip, partly because I only had a few pounds in all the world, and partly because whenever I plan a three-day visit to a port it always extends into as many weeks.

The rugged, slim finger of Wilson's Promontory points downward from southern Australia towards Tasmania, almost closing the eastern exit of Bass Strait, and ten miles from the tip on the east coast of the Promontory is a tiny enclosed bay. That is Refuge Cove, and that is where I planned to rest if need be.

The winds became inconsistent when I was still 300 miles from Cape Otway, the western entry to the Strait on the main Australian coast. That dawn the star Sirius, which I had followed all the way from England, showed as I had never seen it before, flashing vivid reds, blues, emeralds, and yellows, in a beautiful display.

'And what,' I wondered, 'goes on up there?' Is this great Universe, of which this world and Sirius are but atoms, made and held together just for our benefit and destruction? We people are probably nothing more than germs in the blood-stream of some colossal being, and when we let off an atomic bomb he feels a twinge and says to himself, 'I must be getting old.'

Another 100 miles passed and my diary records the first ship seen for almost 2000 miles:

> I passed a ship last night about 11 p.m., probably on its way from Hobart to make Kangaroo Island its departure for the west. It crossed about five miles ahead of me, a blaze of lights, and probably quite unaware of being watched by me, travelling without lights. It always gives a strange feeling, this passing a

ship at night—my world in solitude and proclaimed insecurity, yet so safe in the harmony of *Sheila* in the wind; and this other, a number of humans all in their allotted tasks crowded into the blatant light and noise of a steel hull, from which no Sirius can be seen; driven against Nature's forces without thought as to why God directs the winds to blow or the sea to heave. It seems like blasphemy. What of the creatures inside? Their warm beds, cooked meals, dry clothes, and pay at the end of the voyage— do they know whose hand is at the helm, where he takes them, or care?

After another twenty-hour burst at the helm came a calm grey morning with hardly a breath of wind, but later it came from astern. It's always the same with a good wind; I'd say to myself: 'Now if this wind holds I'll be through Bass Strait in three days from now—' but it doesn't hold. Which is just as well, because after averaging eighteen hours a day at the helm during a week of good winds, I'd be near dead with fatigue and in no fit state to be alone on a boat at sea.

Sights that day put me forty miles from Cape Otway, but there was a thick haze to the north and no land showed before dark. I had an early tea, the wind came with the night and soon after I picked up Cape Otway light, winking its rhythmic message through the darkness straight before the bowsprit. It never ceases to amaze me, the simplicity with which it is possible to wander about in a huge watery waste for nearly five weeks, and come out of it exactly where you want to without (in my case) really knowing why.

I sailed through the night until 4 a.m. and then slept with a light hung in the rigging. From the tone of my diary the sleep must have done me good:

> Slept till 8 and awoke to the most beautiful morning since Creation. Gentle wind, mild deep blue sea, warm sunshine—just heaven after the rain and wetness of the last few days. Put the radio in the sun to dry out, had a huge cup of tea, put on full sail,

and the good wind is such that *Sheila* sails on course unattended, so I'll catch up on sleep—wonderful to know she is taking me home while I do so. Still no ships, amazing, but I suppose they all go into Melbourne; there'll be a crowd in the bottle-neck of the eastern exit.

Before sleeping that dawn I had seen the glow of Melbourne's lights in the sky, although they must have been ninety miles away.

I sailed all that night and all the next day (Friday the 13th!) in an attempt to clear the eastern exit before dark. The weather had steadily worsened during the day and continued to do so until at dusk I was off the Citadel, a rocky island five miles short of the exit. I was very tired, but a westerly gale was ideal to hurry me through, and anyway sleep in that bottle-neck was unthinkable—or rather I couldn't stop thinking of it but it could not be allowed.

I flew through, keeping a check by continuous compass bearings to the lights, because the strong cross-tides can carry a slow ship swiftly off-course on to several unlighted rocky islands which line the route. And I was extra careful, knowing how tired I was. I had planned to signal the lighthouse in passing to report my passage, but in the heavy tide-rip off the end of the promontory huge seas broke over the stern and *Sheila* became almost unmanageable. It was thrilling to enter the Tasman Sea, the same that lapped New Zealand's shores. I turned north towards Refuge Cove and lay off and on for the remainder of the night in the shelter of the Promontory. That evening I wrote up my diary:

> I ran in to where I judged the Cove to be in the early hours, and dawn came just before entering—a narrow hidden entrance in the granite bluffs which I'd never have found in the dark. The Bay itself is a long oval with a sandy beach and its own stream at each end. Hills up to about 2,000 feet rise steeply out of the bay, covered in scrub and tall trees, and made up as far as I can see of huge granite boulders.

> The peace of coming in here after the sea—water like glass, the smell of rain-soaked bush, calls of birds, and the rising sunlight finding its way down the hills to the beaches.

In spite of having had no sleep for two nights and a day I tidied up hurriedly, had a cup of tea, and went ashore to tread again on land and to satisfy curiosity. There was a small tidal stream which seemed as if it would be fresh at the low ebb, a flat area of scrub just behind the beach, and at either end the hills in tall bush and piles of great smooth boulders. On these were painted the names of many ships, probably those going westwards who had sheltered while an unfavourable gale blew itself out in the Strait. I did not add *Sheila*'s name.

On the soft sand of the beach were the marks of a kangaroo followed closely by those of a baby roo (or 'joey' as the Australians call them), and over both were those of a dog, probably a dingo. The stream was still muddy where the tracks had crossed, and on the other side the big roo had turned in defence. The groove made by the heavy tail was plainly imprinted in the soft sand, and the dog marks were dug deep, where it had come up sharp on all fours before the roo's defence. After that there were no more baby roo marks; I cast around for signs of fur and blood but there were none, so I told myself the story that the joey had popped safely into its mother's pouch and escaped with her.

After four hours' sleep I rowed over to the other beach and found that the stream there was crystal clear. There were also the burnt out remains of an old camp, but I could find no sign of the track which I had been told started from there to the settlement the other side of the promontory.

In the evening I wrote:

> The sun has gone to leave the bay very quiet and still; it is so very beautiful with the silence and solitude to hear it. Why do people herd themselves into cinemas, why destroy silence with bands, radio, talk, when this silence tells so much more than any sermon?

Its stillness reminded me of that tropical lagoon where I had anchored after entering the reefs in darkness, and where an early morning squall had almost driven *Sheila* on to the coral. Here she was very close to the beach to make an easier job of filling her water tanks, and so I put down the second anchor, adding in my diary, '... because it's safest in any anchorage, as I've not seen a safe one yet.' And so having done all in my power to make *Sheila* safe for any emergency, I went to sleep content that she was.

39

Dismal weather—the drake—meditation—departure

The squalls struck in the early hours of the morning, tearing down the gullies to slew *Sheila* on her anchors and force her heeling over as if heavily pressed under sail. With the dawn came rain and bitter cold, the clouds flew overhead, and the radio reported very rough seas and a blizzard in the Alps only forty miles away. I would have sailed at once as the storm was going my way, and to save eating precious supplies while getting nowhere, but there was not one dry stitch of clothing nor a dry blanket on board. The thought of being out there in nights of bitter storm kept me waiting unhappily for the sunshine.

The storm continued and Sunday brought only some of the peace for which I'd been so grateful on others:

> Noon—rain stopped for a bit so went ashore, washed salt out of sails and heated up three gallons of fresh water for a bath—ended with most of the water in a warm soapy deluge all over me, and sheer heaven after the weeks of cold salt water doing so. Then I pelted naked over the sand, feeling strangely immodest but wonderfully free, to the deepest part of the stream and submerged wholly in the ice-cold water. Isn't it the Swedes who roll in snow after a hot bath?—which must be even more wonderful than this.
>
> Later—it's the hell of a night, heavy squalls of rain, vivid lightning, and tremendous rolls of thunder. The gusts before dark were racing over the bay, lifting swirling white clouds of spray before them, 50, 100 feet into the air. Now they are invisible but I can hear them approaching from far up in the hills, as they come

screaming down the gullies and then batter *Sheila*. I bet she's as pleased as I am to be here and not 'out there'. Gale warning on the radio again, another blizzard… I wish it wasn't so bloody cold.

The next afternoon, with the hatches closed against the penetrating rain, I stood gloomily in the cabin looking at the dismal beach through a porthole. And at the far end of the beach was a fat, complacent duck, pecking idly in the sand and apparently quite unaware of the reason for his existence. I however was hungry and short of food.

I hate killing anything, but the argument between Hunger and Conscience became so intense I left it for them to decide the issue.

'Dash it all,' said Hunger, 'the duck's there to be eaten, and he can't eat it alive.'

'He's not starving,' replied Conscience. 'And anyway by killing this duck he'll stop the birth of many others, which might save people who are starving.'

'He's starving enough and has a hard trip in front of him,' persisted Hunger. 'And it's making him so bad tempered he'll soon feel no affinity with you anyway.'

'All right, you force me to a concession,' said Conscience. 'I'll direct the duck to meet this situation, and you direct Hayter. If you're strong enough he'll get the duck, and if you're not he won't.' And they came to me with this decision which I accepted.

I changed into a sodden khaki shirt and shorts and rowed to the other end of the beach exuding goodwill to all Nature, hoping the duck would feel it. Taking an oar I went about 100 yards into the bush, circled, and made a careful way through the dense tangle towards the duck; for the last fifty yards I crawled full length with patient caution, hoping the caution to surprise a duck wouldn't surprise a snake.

There was the duck only a few yards from the edge of the jungle and I tried hard to stop my mind from concentrating on him, because this creates a telepathy to which animals are particularly susceptible. I could feel a current between me and the duck, and I could feel his uneasiness come back to me. He took to the water and paddled around about twenty yards out, uncertain of his alarm, and I waited motionless for twenty minutes while he convinced himself he'd been imagining things. Then he landed again about thirty yards along the beach.

I backed into the bush, circled carefully, and reached the edge of the cover only five yards from him. As he had taken to the water rather than to flight, I thought he was probably one of those birds which require a bit of take-off and rise slowly. My plan therefore was to charge out, fling the oar, and grab him in a flat-out racing dive—on to the sea or the soft sand as his reaction demanded.

Waiting until he was busy pecking at something I leapt out, raised the oar—but it never left my hand. The duck simply opened his wings and rose vertically, circled twice, and disappeared over the hill. Hunger said, 'You silly fool,' and Conscience, 'Serves you right.' So that was that.

The wintry storm continued to rage, penning me inside the damp cabin with my sodden clothes and blankets, depleting my meagre supplies without adding compensating miles towards home. I was bitterly cold, and out of the enforced idleness grew boredom, and boredom is hell; and hell is loneliness, when we place ourselves in a separate existence from forces we cannot change. And so, exhausted perhaps by idleness, I accepted the storm and the imprisonment it imposed, and worked long hours in mental fields where there is ever work to be done.

By this time towards the end of that week the hours were far into the night, and the small hurricane lamp on the saloon table was the one source of light and warmth in the vast darkness in which *Sheila* lay, held secure by her anchors while I forgot her in my thoughts. In those hours the first gleam of their full significance came to me—so many experiences and thoughts in the past had conflicted, like the scattered pieces of a jig-saw puzzle; but now each piece fitted neatly into place, as one truth confirmed another, and only then did the picture of the whole begin to take shape and make sense. My mental voyage too was almost done.

The oil in the lamp ran dry in the early hours of the morning, leaving me with thoughts not seen so clearly then, so I lost their devastation in sleep.

The storm continued the next day, and I decided to go ashore and seek again for the path I had been told ran across the promontory to the settlement the other side; my few pounds converted into supplies would make a big difference to the voyage ahead. Short rations were one reason why I felt the cold so much that it drove me from the helm, thus prolonging the voyage and imposing an even stricter rationing. I also wanted to send a cable home to relieve the cruel anxiety of those who suffered more during my voyage than I.

Entering the bush at one end of the beach I made a wide circle inland to come out at the other, so as to cross any path leading out of the area. There was no sign of a path. I still could have found my way across the ten miles to the settlement, by compass and by reading the contour lines marked on my chart, but it was not worth the risk—it would have been easy to slip and break a leg among the huge haphazard vine-covered boulders, or get bitten by snakes with which the area crawled; nor did I like to leave *Sheila* for so long while the tremendous gusts gave her anchors no rest.

The wind changed on the tenth morning after arrival in Refuge Cove, the overcast broke into clumpy cumulus, and the radio reported gale moderating. So that day I dried out what was possible between the showers of fine drizzle, took *Sheila* over to the other beach and filled her water tanks, and made all ready to sail.

There were a bare two weeks' supplies on board, at the normal ration scale, and I thought of sailing the ten miles down the coast to the lighthouse for more, but doubted whether I would be able to land with the surge running. This was the last leg, the very, very last, and even a ten mile detour from a 1200-mile crossing seemed to be out of place.

40

Adverse winds—emergency ration scale—the cyclone—land-fall— the handkerchief— the Westport Bar—home

Early on the morning of 23rd April I hauled in the two anchors and drifted out of Refuge Cove on puffs of wind; to leave such a place under sail seemed to be more in harmony with all I'd found there. The high granite ridge of the promontory sheltered the water outside, until about five miles offshore a steady westerly picked me up and *Sheila*'s clean bow cleaved a way direct for home.

A week later sights confirmed that I had covered 420 miles, or a third of the distance to New Zealand. The winds had been cold and bitter, and leaking decks had soaked everything below once more; drifting acres of rain and flying spray kept me chilled to the bone, unprotected by the useless oilskins. I had re-sewn the seams, but the strong sail twine pulled out of the soft water-proofing material.

Then the wind turned east, dead ahead, against which *Sheila*'s Geraldton-cut mainsail was almost useless. This meant that I dropped below the distance-made, rations-allowed estimate, and so immediately adopted the emergency ration scale.

The adverse wind blew at half a gale for eleven days, and I arranged my three daily meals around the clock at eight-hour intervals. Because of short rations every mile to the east was gold, so I sailed every hour of the day and night with only the barest snatches of sleep, and in the eleven days covered hundreds of miles to the north and south but only gained eighty to the east. It was a lot of work for so little, but as it turned out those few miles probably made all the difference between the success and failure of the whole voyage. Thus we may never know the value of the things we do, however small the gain or great the payment at the time.

The days were wild and wet, with low-flying cloud and curtains of rain; the nights were black as ink and bitterly cold, with what the Australians call a lazy wind. It's too lazy to go around you. Hunger was growing and greatly increased the effect of cold, when the mental reaction becomes worse than the physical. You feel despondent.

There seems to be a limit to what we are asked to bear, and perhaps when a thing becomes too much it simply topples over like a wave and levels off for a time. This seems to come about through some protective agency beyond our influence, but I'm not so sure of that. My diary enters this event:

> Turned on the radio just before 8.30 p.m. (7 a.m. GMT) for the first time to Radio N.Z. to get the Time Signal. And out of the radio came the clear notes of the Bell Bird—it brought back all my deep love of this country; I've said I'd never be able to live there and criticise it violently, but only because I love it so much. The Bell Bird brought back memories of earlier days. The purity of the sound seems to pervade all around it, including oneself, which is a very nice feeling. I'm going to make my home in New Zealand—it's the way I feel now anyway.

And for my own conversation I added, 'You smug ape, you'd make your home anywhere now.'

There used to be a lot of bell birds in a patch of bush just behind the house, and I knew they'd still be there. I knew also that the home and the fire-side (to which my thoughts returned most vividly then) would be just the same. My father usually took a big rumpled chair on the far side, while my mother sat immediately in front near the table with a pile of socks and mending on it. And others of us had taken the other side opposite my father, to talk, read, or just stare into the fire. So many dreams had been born in that fire, perhaps after reading a stirring tale of the North-west Frontier, of sailing to far places and of storms at sea; and so many had come true.

The crowded mantelpiece too would be the same—my gymnastics cup, some cheap bronze medals we kids had won in various sports and borne home with such pride, all so unimportant, all so valued. And I felt sure a few opened letters would be pushed behind the clock—those were letters which anyone in the family could read. We had a 'thing' in our family about never reading anyone else's letters, and those left anywhere but behind the clock were as safe as in a vault.

Fair winds came with the bell bird's call and I gained another 200 miles to the east—and then came near-disaster.

The wind was mild in the morning, but during the afternoon a huge fan of cirrus ('mares' tails') and cirro-cumulus clouds spread outwards from the north, and a heavy swell came in from the same direction. The sun was cast in a livid, unhealthy haze. Around it ran a steel-blue halo, and when it set in the evening it looked like a ripe carbuncle. The weather steadily worsened during the night and all the next day, and the swell increased. These were all the text-book signs of a coming cyclone, and because the bearing of the wind did not change I knew that I was right in its path.

A cyclone is like a gigantic whirlpool of air with winds intensifying towards the centre, only in the 'eye' of the storm there is no wind but terrific seas, as great waves aroused by the winds on all sides converge to fling themselves madly against one another. It is mainly in this eye that the danger lies and it must be avoided at all costs.

The Pilot Book said of these storms, '...and the high confused seas near the centre may cause considerable damage to large and well-found ships, while small vessels (for example, destroyers) have foundered.' It didn't even mention *Sheila*! However because the path of these storms can be roughly assessed by the direction of the wind, it is possible to take action to avoid it if you have a fast enough ship; in a sailing ship the procedure is a little more complicated, and in my case meant scuttling away off course. I simply didn't have enough rations for such diversions, and anyway I knew that a cyclone in latitude 40 was unlikely to carry the tremendous forces of those met in more tropical latitudes. So I decided to hold my course, having more confidence in *Sheila* to ride out a cyclone than in myself to live on nothing.

By dusk on the second day it was blowing gale force, and still increasing. *Sheila* lay hove to under storm canvas, but at nine that night I struggled to take it in before the still-rising wind tore it to shreds or pulled out the stick. *Sheila* felt easier thereafter, steadied (without being over-burdened) by the tremendous wind in her rigging, and the only discomfort came with occasional breakers against which no small ship has protection.

Dawn came slowly through the dense rain, but at 10 a.m. the screaming wind stopped suddenly, the sky cleared overhead into bright sunshine, and we lay in the centre of the cyclone, enclosed by a huge beyond-horizon-wide circle of black clouds under which the storm still raged. There were only faint puffs of wind, but the seas were gigantic, rushing into each other, lifting into tall top-heavy triangles and flopping back to cause more trouble. I could do nothing to steady *Sheila* without wind, the un-rhythmic tossing and battering placing terrible strains upon her; during that day the eye of the storm slowly passed over us and the other rim approached.

We entered the other side just after dark that night, and flew east before the screaming rush under bare poles. It eased to gale force near midnight when I got on storm canvas, and replaced this just before dawn (when the wind eased further) by the closely reefed main and storm jib. I reckoned we covered 100 miles in the next fifteen hours, and at times when *Sheila* was lifted high on a crest and the full force of the wind hit her, she was flung far over on her side until half the sail was flat in the water. It was bad seamanship to sail so hard in such weather; the short rations impelled it, but it imposed the great strains and discomfort that bad seamanship always does. And so I had my cyclone.

The ration situation was poor, and just over half-way across the Tasman I had seven 12oz. tins of meat and six 10oz. tins of vegetables left. I'd been on the emergency scale for the last two weeks, one tin a day alternate meat and veg. to give variety, and only once did I succumb to the obvious common sense of mixing a tin of each together into a more pleasant stew, as a two-day ration. It is too difficult when hungry to face a lovely hot pot of such delicious stew, enough for the one good meal you need, and instead ladle out just one-sixth to draw it out into the full two-day ration it had to be. My meal times were 8 a.m., 4 p.m., and midnight, and there is a pitiful wail in my diary after the storm:

> Left the helm soon after dawn, and was so ravenous thought I'd have my breakfast then at 6 instead of waiting until 8. It was truly delicious, but mugger me, it's now 10 hours before I can have another.

Followed a few days of winds from everywhere, all unpersistent and unreliable, in which I made what easting was possible, but in the fifth week out came a strong steady sou'-westerly. It had all the feel of staying for several days, and with only 300 miles to go I determined to stay at the helm either until it blew out or until I reached New Zealand.

My object was to make a landfall at Farewell Spit and get inside its protective arm, where in the comparatively sheltered waters of Tasman Bay my engine would take me the remaining forty miles to Nelson if the wind dropped. Home—I just couldn't believe it; and yet to my surprise the idea of leaving that horrible life on salt water was tinged with sadness.

That wind held for three days. I know there was little time off for meals, such as they were; entries in the log are merely 'sailed till dawn', and scrawled haphazard across the page whichever way the book had opened; and there is no mention of sleep, but I remember I did go below one night for three hours, numb with cold.

The night of 24 May was clear and moonlit. The main coast or even Aotea-roa (the long white cloud) was not visible before dark, but by midnight I knew I must be close. At times a dim invisible shape appeared in the distant moonlight on the starboard bow, so invisible it couldn't be New Zealand, yet so dim it could have been something. I wasn't sure.

The wind dropped two hours later, at 2 a.m., releasing me from the helm after an almost unbroken three days and nights, so I went below to catch up on sleep before more wind recalled me.

A beautiful dawn came, the sea smooth, the sun warm, and the rugged bush-clad coast of New Zealand lay twenty miles to the south-east; the low-lying sand-hills of Farewell Spit lay about forty miles dead ahead, but still under the horizon. There was one meagre meal left on board, six ounces of tinned peas, and as many cups of tea as I wanted without sugar or milk. So I sat on deck in the sun and dried out clothing, content with the world as it was because I couldn't do a thing to change it. The warmth and the stillness were heaven anyway.

There was enough petrol on board to carry *Sheila* as far as Farewell Spit in calm water, but what if the wind got up from ahead after thirty miles to leave me ten miles short? It could then take another two days to get inside the spit, and another two to Nelson. Hunger in itself is no great danger for a few days, just unpleasant, but the fatigue and weakness are dangerous. It only needs a slightly careless hand-hold or one without quite enough strength in it to let you fall overboard, to mention only one of the myriad forms weakness can take.

The alternative was to move in towards the coast and make Karamea, the nearest port and no further away than the spit. It wasn't Nelson, but if the wind did go south after all, I could use it to make the spit and still have my petrol intact; and if the wind came from ahead, with food in me I could take *Sheila* to Nelson later.

The wind came at noon, straight off the spit, so I got on sail and headed for Karamea. I had no large-scale chart or Pilot Book of that coast (stupid economies), and only remembered that at school someone who came from Karamea had mentioned that ships called there for coal; and if ships could enter so could *Sheila*. On arrival there that evening I discovered that over the years since I had left school the river had silted up the bar, and nothing could get in.

There was a big surge on the beach, too big to land in my pram, and although people were walking about within sight on shore I knew no one could get out to me. The bottom was good holding ground so I anchored for the night outside the breakers, had my last meal, and slept for twelve hours. The alternatives for the morrow were either to find a sheltered bay with a homestead or shell-fish, or go another eighty miles down the coast to Westport, which had an established harbour.

Next morning, with my goal of years only a few hundred yards away—but with the breakers between, it might as well have been as many miles—I sailed south down the coast thinking of bacon and eggs. After about fifteen miles I passed a long spur running out into the sea and looking as if it might have a sheltered beach inside it, so I anchored just within the point with the idea of going ashore in the pram and collecting a sack of mussels. The beach was a mass of broken rocks on which the heavy surf broke angrily, and under the drizzle of cold rain it all looked very unfavourable, but suddenly I felt an over-powering urge to get ashore at any price.

My mind supported this idea as being the wisest course—Westport still lay a long way to the south, and the favourable wind might drop at any time to leave me in a worse situation than the first, out of petrol range of anywhere. But deep down was a small still voice, so far away I could pretend not to hear it but I couldn't still it.

'You're in a panic,' it said.

I laid out the second anchor, closed the hatches, and unlashed the dinghy. It was poised ready to go over the side but this voice, this feeling, made me pause; it was the same feeling of premonition that had stopped me plunging into the sunlit Mediterranean that morning when a shark followed astern without my knowing it then; it can also (I've found) mean I've forgotten something, like a door-key, and it always pays to pause.

'Of course,' I said to myself. 'The hanky.'

I pulled the pram back inboard, threw open the hatch, went below to the locker and took the child's hanky out from between the thickly-folded towel, still neatly pressed and dry. It made me smile to remember her serious gaze as she had stood before me.

'Do you promise?' and I had said, 'Yes, I promise,' and trusting me she had given her gift. It was warmer in the cabin than in the rain outside, so I decided to have a final cup of tea before trying to get ashore, because it seemed highly probable that more than the rain was to wet me.

And while I drank that tea I took the opportunity to face that voice, because it still persisted. It said plainly that my chances of landing that light pram in that sea on those rocks, without damaging it or myself so much as to make return to *Sheila* impossible, were slight. And if the weather got worse *Sheila*'s anchors would not hold her.

'You're not going ashore for food,' the stillness said. 'You're in a panic and deserting your ship, and I'll ruddy well see you never forget it whatever stories you may hatch up, whatever fame the headlines give, whatever sympathy you receive.'

I had been on the look-out for some reaction to hunger and fatigue, and it is always near the end that distractions become strongest, but I had not expected them in this form.

So I finished my tea, put the hanky safely back in the folded towel, pulled the pram right inboard, up-turned it over the skylight and lashed it down, took in both anchors, and sailed for Westport.

It is not possible to state that the child's handkerchief and trust saved me perhaps from disaster, certainly from a decision I would have regretted for the rest of my life; but it is equally impossible to deny it. Isn't this the child's fairy-tale, when a gesture of simple trust beyond any thought of reward turns the lowly beggar boy into a handsome Prince; when a gentle kiss beyond desire awakens the Sleeping Princess and banishes the ugly witch under whose spell she lay; and the Prince and Princess live happily ever after? It's when we grow up and lay these stories aside that we lay aside their miracles too.

I arrived off Westport at three in the morning. Knowing the type of coast and that a river flowed through the harbour, I knew there'd be a bar across the entrance and that with the big swell it would be dangerous. The red and green entry lights were clearly visible, probably on the end of the breakwaters, but the red light seemed to be to starboard and the green light to port, entering.

This is in accordance with the old system which changed soon after I had left England, and it seemed that Westport had not effected the change, which is opposite to the old; but I could not be sure. It's hard to explain the tricks that lights and angles can play at night, without a chart and not having seen the place in daylight—those lights could still be right, only appearing different owing to some angle in the breakwater I could not see; like the entrance lights into Bône which could have wrecked me.

It was worth having a closer look so I eased *Sheila* in under sail; under sail because when the wind is steady and dangers lie close to leeward, sail is far safer than power and the ship handles better. Under power a drop of water or a bit of dirt in the carburettor jets, or some temperamental nonsense in the electrical system, can leave you bereft; and under sail you can feel what a ship tells you, which is often what your eyes, ears, or even a chart cannot. The feel of *Sheila* to the waves, a feel words cannot explain, told me the situation was lethal; and being uncertain of the entrance lights I applied a golden rule—when in doubt get away to hell out of it, back to open water.

I headed out about a mile, hove to, and stayed awake until dawn in case wind or current took me into danger. It was not wise to go further out to sea to sleep safely, in case a hard land-breeze came with the dawn and made it difficult to regain the port.

The drizzly dawn came slowly and I took *Sheila* close inshore to clarify the entrance. The river flowed strongly out of the harbour, its strength contained between two long breakwaters reaching out towards me, whereafter the out-flowing water piled itself against the heavy incoming swell to form an area of high-reaching top-heavy breakers such as never happen in their normal state. So there again was my goal for the past six hard years and 18,000 miles, with such breakers in between that I would normally never have considered entering.

The longer I delayed the less likely would I be to make wise decisions, and the less strength with which to carry them out—*Sheila* needed my best as much as I needed hers. The weather might get better or it might get worse, when the bar would become unthinkable, and I had no food to stand off several days waiting for an improvement. The attempt had to be made and the sooner the better, but first I went below to make a last cup of tea, to consider all the factors that might arise and make the most careful plan for dealing with them.

The greatest danger of the breakers was not their size but their speed.

Perhaps many have seen what happens to a light dinghy coming in to a beach on a surf— it is gripped by the forward edge of the wave and held in a speed which makes it uncontrollable, and the slightest deviation from off-straight forbids correction until the dinghy swings broad-side on to the wave, is toppled over and submerged. That was exactly the danger to *Sheila*, designed for a maximum eight knots, but those in-rolling seas were travelling at about thirty. And near the breakwaters, even if I could delay the broach-to, I would not be able to control direction, and to hit either of the breakwaters at thirty knots would be disastrous. And no swimmer could have lasted more than a few minutes in those seas and currents—I reckoned, in cool assessment, that I had a fifty-fifty chance of survival, and it was a bitter thought that in a few minutes it might all be over on the very threshold of my home.

I stripped off to shirt and shorts in case of a swim, because you never really know your chances and I'd certainly have no hope in the heavy clothes I was wearing. I then put two reefs in the mainsail, to give *Sheila* with that wind just enough power to drive her against the current with the minimum speed. The mizzen was furled in case a sudden gust swung her stern off course. I started and tested the engine, and let it run in neutral so it would give no extra speed and no spinning propellor would cause turbulence astern to upset a coming sea, but so it was ready for use once the shelter of the breakwaters had been reached. The wind was less there, and the two-reefed main would not have held *Sheila* against the out-going current; nor in that confined space would I be able to leave the helm to put on the extra sail needed.

It would have been normal seamanship to have trailed a heavy rope over the stern, so that its drag would lessen the chances of *Sheila* being carried forward with the speed of the waves; but the following seas might have swept it forward and around the propeller, to foul it and stall the engine when I did put it in gear. I had to be sure of that extra power when I lost the wind, or else be swept back into the breakers really out of control.

Then I again took the child's hanky from the locker, and buttoned it into the pocket of my shorts, so that whether I stepped ashore or my body was washed ashore trust itself would not be broken.

It was a very strange feeling deliberately and with calculation entering a situation which I hardly hoped to survive. For fleeting moments I knew it was to be death, but underlying, far down in my consciousness was a smooth, strong tide which I sensed was immune to all things, even death. And not to sense that tide was to panic, but that was mine to control and my only prayer was for the strength to do so.

All was ready. I lashed myself to the helm with a slip knot, closed all hatches firmly (with my warm clothes just inside in case I was to need them later), and turned *Sheila* towards the entrance and the breakers, placing her so that her bow pointed dead straight before the waves and into the dead centre of the entrance—and noted the compass course. Then I took her right to the very edge of the outermost breaker, keeping that compass course as accurately as possible; just before we got to the breakers her bow pointed 200 yards north of the northern breakwater.

Sheila answered as she always does under sail, turning as she lifted high on a big sea which broke with a roar only yards beyond, and I took her back into deep water to the distance of the original starting point.

Then I sailed south until the same compass bearing bore to a point ashore 200 yards south of the southern breakwater, plus a bit more. There was a drift up the coast which I had to allow for, because if I started from the wrong point and tried to correct on the way in, it would throw her off the dead straight before the seas at speed and she would become uncontrollable.

So in we went, and in seconds entered the breakers. *Sheila* lifted, surged forward quite happily, the broken crest climbed over the afterdeck and spilled into the cockpit. She rode three waves beautifully until a monster bore down, so huge I felt there was no hope of survival, knowing that this one wave would decide the issue of the whole voyage. If I could maintain a dead straight course before it, it would carry us through the turmoil to the safety beyond, but if we wavered one iota, the very strength that could carry us through would annihilate us, either swinging us broadside on or into one of the breakwaters.

This was known as even then *Sheila*'s stern lifted, and I forgot the breakwaters, the entrance, all fear vanquished by inevitability as I pinpointed my mind into the helm. With this I could no longer steer the course I myself wanted, but only maintain the course it was disaster to evade.

Sheila became gripped in speed, her hull sucked down in the tide to deck-level; her bow wave was two solid masses of water, hurled aside so high I could see nothing beyond them, not even the light structures on the ends of the breakwaters. I could feel her balanced on the helm, which needed only a finger-tip to maintain with occasional gentle strength to correct tendencies about to develop; if these had developed, the strength of ten would not have saved her.

I was only half aware of the wonderful scene through which we roared—huge tangled masses of tortured water, glimpses of smooth-welling eddies, terrific noise, and through the soles of my bare feet I could feel *Sheila* vibrating in every plank. Never have I felt so one with her, bringing that glorious surge of pure exhilaration and thrilling experience when you make powerful destructive forces not your enemy by mastery over them, but your own glory by attaining harmony with them.

Then the crumbling wave returned to its own, its destiny fulfilled, leaving *Sheila* in the centre of the channel between the breakwaters, balanced peacefully between the current from the land and the wind from the sea. I eased the gear into ahead, and as we moved sedately up the placid channel I pushed open the hatch and reached for my trousers.

'Thank God,' I breathed, 'I didn't try to land up the coast.'

Soon the channel widened and bifurcated, one branch leading upstream to the wharves, and one to a sheltered lagoon in which small boats lay. I put the engine out of gear and took in sail, feeling every fold and reef point known so well, and going into gear entered the still harbour which lay under the gentle, cool mist of a Sunday morning.

I again paused in the middle of the lagoon to look about and decide where to anchor or tie up alongside, and on the iron deck of an old dredge was a figure muffled against the morning cold. He waved me in, and as *Sheila* drifted alongside he took my ropes and made fast.

With his kindly welcome and congratulations came my own reaction, and I climbed up the iron side of the dredge, crossed a gang-plank on to New Zealand soil, as in a dream. My obligations to the Customs, Immigration, Port Health and others never entered my head.

My new friend, this stranger whom I somehow felt I'd known for years, took me to his nearby home. His family were out at early morning Church, so he busied himself getting breakfast, pulling out dry clothes, stoking up the range for a hot bath, laying out a fresh towel and his shaving gear; and then he said, 'You'd better have a brandy,' and pouring out a medium tot he handed me the glass. I drank it slowly, and it made me feel more in a dream than I was already.

'I won't give you another,' he said. 'There's plenty there, but a cup of good hot tea will do you more good now.' And I think I was more grateful for the understanding of that denial than for anything else he gave me.

And soon I was on the phone to my people, to banish the fears they had borne unspoken for so long. And then the port authorities were on the phone to me.

The Customs were most kind, accepting my apologies for what was officially a grave transgression. The Port Health Officer collected me in his car, and on the way to his office a frantic desire for something sweet swept over me—I'd had nothing with sugar in it for weeks.

'Stop at a sweet shop, please,' I said, 'even if I have to break it open.' But one was open and I rushed inside, only to return empty-handed.

'Don't you want it after all?' he asked.

'Yes,' I said, 'I'm longing for it but I've got no local cash.'

So this official went in and brought out two thick slabs of the most wonderful chocolate I've ever known. The brand was unimportant.

Others came. My old school Nelson (I don't know how the news spread so fast) rang a local Old Boy, who came to offer anything he could and gave me a great deal of help. A fisherman, who looked as if he might be a bit of a wag normally, came and offered to look after *Sheila* if I intended going away for a brief visit home. It is asking too big a responsibility to leave *Sheila* in someone's direct charge, but he accepted it.

(When I returned a week later to take *Sheila* on to Nelson, I found he and others had cleaned and dried her out, taken my salt-soaked blankets ashore, and his wife had washed and ironed my clothes. They cannot know what all that meant to me.)

I told my story to the local press. The editor took me back to lunch, and I found that his wife had rushed around the town to borrow a good steak, because she thought that would be best for me. And that paper wrote none of that exaggerated nonsense which to me is always a desecration. I know better than anyone else to whom the real credit for my safety lies, and to misplace that credit seems to include me in a claim I would not dare to make.

That evening my young brother and a friend arrived by car, having driven through from home; and the next evening the car drew up at the white gate at our garden. I tried to get out as if I'd only been for a drive into the village, and my parents came out with loads of Hayter reserve, their grins almost getting out of control into tears, which didn't matter because we all knew they were there.

Then we went inside. The room was just the same, the fire-side and the mantelpiece and the letters behind the clock; and I knew I'd never been away from the home they had kept so safe.

TECHNICAL GLOSSARY
(Refer to drawing opposite)

1. Mizzen mast & sail
2. Bumkin
3. Mainsail
4. Topsail (*Sheila* may not have had one on voyage)
5. Gaff (spar holding up mainsail)
6. Headsails; jib (outer), staysail (inner)
7. Forestay
8. Bowsprit
9. Bobstay
10. Shrouds (wires supporting the mast)
11. Hounds (upper fixing point for shrouds)
12. Chainplates (lower fixings for shrouds)
13. Main boom
14. Topping lift (supports boom when hoisting sail, then slackened)

Sheets are the ropes controlling the lower rear ends of each sail; the *clew* is the lower rear corner of the sail; a *cringle* is a reinforcing grommet at the clew to take the strain of the attached sheet (for a headsail), or of the *outhaul* holding the clew towards the aft end of the boom, for the main or mizzen sail. A *genoa* is a large headsail replacing the jib, and reaching back past the mast, for light to moderate wind conditions. A *spinnaker* is a large, very fully cut headsail for downwind use, held out from the mast with the aid of a spinnaker pole.

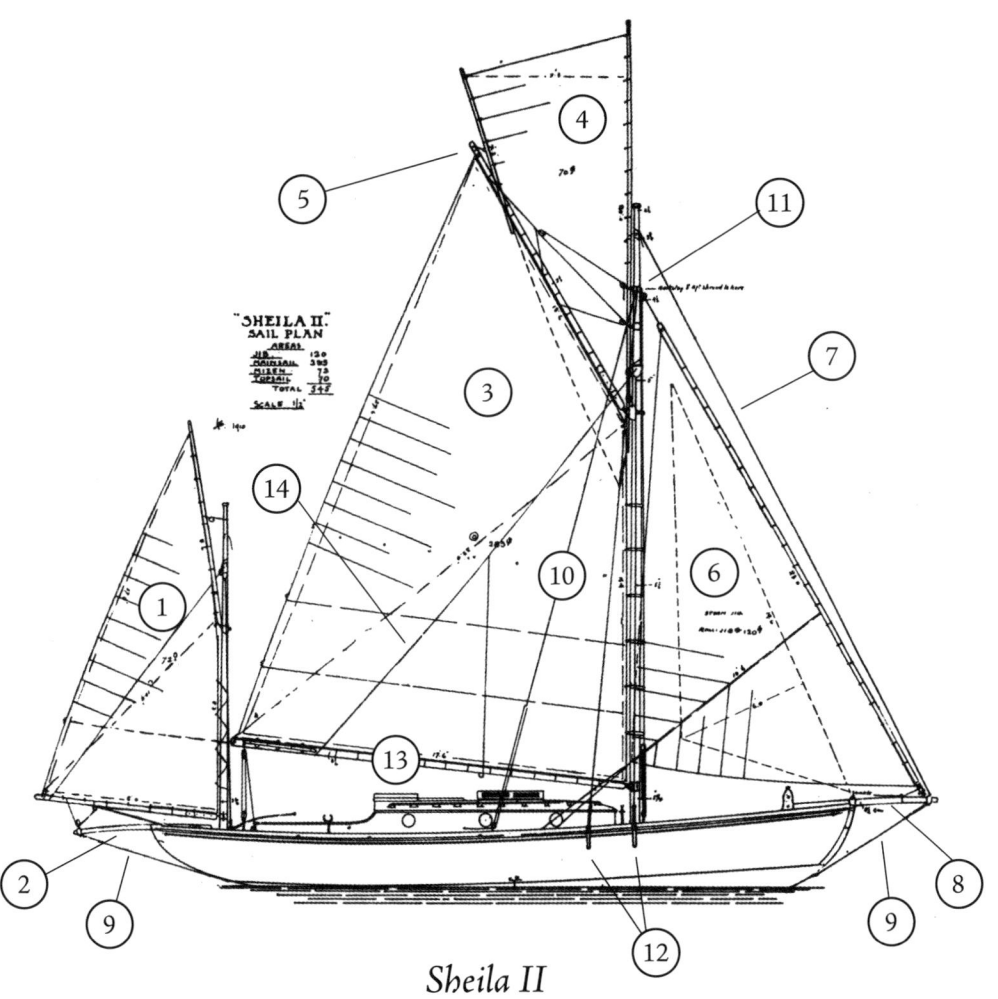

Sheila II

Design No 99 by Albert Strange, 1908,
for the artist Robert Groves
Built 1911 by Dickie, Tarbert

Length, on deck	31ft 7in
Length, waterline	24ft
Beam	8ft 6in
Draught	4ft 11in
Displacement	6tons 2cwt
Sail area	545 sq ft

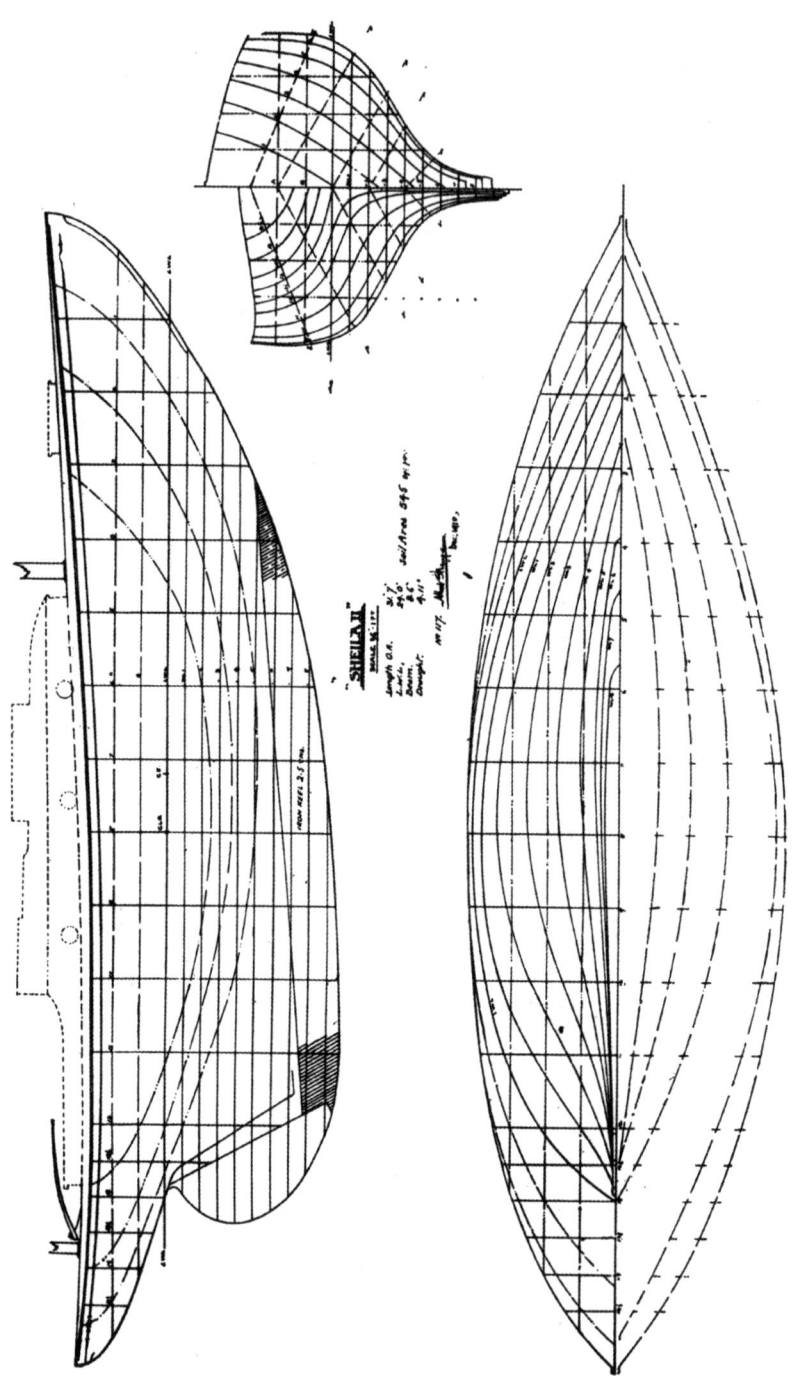

Sheila II: Lines

Sheila II: Accommodation, as drawn by Albert Strange in 1908

INDEX

Abrolhos Islands, 245, 246, 279
Abu Ail Channel, 111
Acts of the Apostles, 185
Aden, 35, 98, 112, 116, 119-20, 165
    ashore at, 120-26
    Gulf of, 128
Admiralty Chart Depots, 46
Alboran Island, 51-55
Algeciras, 48
Algiers, 51, 66, 68, 78
    ashore at, 57-65
    Casino, 63-65
    passage to, 43-51
    Paul, 55
    Yacht Club, 54, 55, 61
Assab, 114, 115
Australia, Western (*see also* Fremantle and Geraldton)
    Australian social life and conditions, 256-7, 258-72, 294-98
    crayfishing in, 279-82, 290-93
    Horrock's Beach, 249-51, 254
    labour conditions in, 258-66, 273-78
    landfall in, 248
    passage to, 226-48
Australian Bight, 273

Bab-el-Mandeb, Straits of, 104, 112
Bali, 220-30
    Bali Hotel, Den Pasir, 220, 222
Bali Strait, 226
Banka Island, 207
Bass Strait, 298, 301, 302
Bateau Plaisir, Cafe du, Bône, 67-8
Bawean, 207-14
Berlenga Isles, 30
Bernier Island, 233
Berthon Boat Company, 21
Biscay, Bay of, 26
Bombay, 117, 129, 140, 169
    Ann and Neil, 149-57
    ashore at, 142, 148-53
    passage to, 133-42
Bône, 55, 60, 61, 71, 317
    ashore at, 63-5
    Bateau Plaisir, Cafe du, 67-8
    Kasbah at, 66-8
    Yacht Club, 63-6
Bougie, Gulf of, 60
Brothers Light, 103
Buddhism, philosophy of, 108, 157, 175, 181
Bull-fight, 41
Butterworth, 195

INDEX

Cabo Aguada, 154
Cabra, s.s. (?), 82
California, yacht, 176, 200
Cannanore, ashore at, 163-9
Cap Bon, 69
Cap Bougaroni, 60
Cap de Fer, 60, 61
Cape Caxine, 50
Cape Comorin, 173
Cape Ferrat, 43, 46
Cape Finisterre, 28, 29
Cape Leeuwin, 299
Cape Otway, 301-2
Cape St. Vincent, 33
Cape Tres Forcas, 44-7
Carnarvon, 233
Ceylon (see also Colombo), 142-3, 173, 175-9
    Burghers, 176
Cheshire, troopship, 110
Chusan, s.s., 110
Cochin, 168-70
    ashore at, 172
Colombo, 179, 182
    ashore at, 175-9
    passage to, 172-4
Communism, theory of, 231, 263
Conze, Edward, 176
Cyclone, 311-12
Cyrenaica, independence of, 83

Dabhol, 139
Daedalus Reef, 103, 105
Darwin, Charles, 87

De Lesseps, Ferdinand, 95
Deep Water and Shoal, 104
Delhi, Great Moghul of, 139
Den Pasir, 220
Derna, passage to, 76-83
    Ali, 83-4
    ashore at, 83-5
Dilly, 200
Djibuti, 117-9
doldrum belt, 177, 180-5, 197, 226
D'Urville Island, 25

Edinburgh, Duke of, 66, 242
Egyptian police, 98-9
Elizabeth, H.R.H. Princess (now H.M. The Queen), 72, 242
Empire Trooper, 122
Etna, Mount, 71

Farewell Spit, 313-4
Farilhocs Isles, 30
Fatma, Isle of, 115
Fratelli Rocks, 69
Fremantle, 226, 229, 233, 234, 241, 242-50, 266, 272, 279, 289, 300
    ashore at, 289-93, 299-300
    crayfishing at, 290-3
    Ted, 291-2
French Foreign Legion, 119

Gasper Strait, 207
'Gateway of Tears,' see Bab-el-Mandeb

Geraldton, 244-89, 309
  Bob, 271-2
  Harry (Customs Officer) 250-8
  Jeff, 259, 288
  labouring at, 258-66, 273-8
  Mick, 258-60
Gerbault, Alain, 93
Gibraltar,
  passage to, 25-35
  ashore at, 34, 41-3, 71
  dockyard, 54
Goa, 149, 154-5, 163
  ashore at, 157-62
  inhabitants of, 155
Gold Mohur, 126-7
Goose Barnacle 230
Gothic, s.s., 242
Gozo, 71
Great Barrier Reef, 205, 244
Gurkha Regiment, 25, 35-6, 42, 122, 158, 163, 177, 191, 201, 268, 280
  Adjutant of, 36, 145-7
  Gurkha Officers' Club, 191

Hamilton, Peter, 201-2
Harrow Meteorological Office, 35
Hillary, Sir Edmund, 226
Hinduism, philosophy of, 86-7, 109-10, 158-9, 162, 188
Horrock's Beach, 249-54
How to Win Friends and Influence People, 203

India (see also Bombay), 54
  hospitality and friendship in, 143-53, 172
  Prohibition in, 142
Indian Navy, 147, 172
Indonesians, 186, 200, 204, 214
  views on United States, 218-9
Ipoh, 191
  Gemma and Peter, 191
Ismailia, 94-6, 134
  Cercle Voiture, 96
  English Club, 97
  French Club, 96

Jabal al Tair, 110
Johore Strait, 200, 202
Jubal, Strait of, 100-1
Jubal Zubair, 110-11
Jurien Bay, 272

Kangaroo Island, 301
Karamea, 314
Kasbah, Bône, 66-8
Khanderi Island, 140
Komodo Island, 200, 204, 215
  'Komodo Dragon,' 204
Kwan, Yin 236, 250

La Galite Channel, 69
La Linea, 40
Life magazine, 59
Lisbon, 30
Lloyd's (*Sheila*'s registration letters), 34

Louvre, the, 220
Lymington, Dorset, 140
   Ship Inn, 22
Lymington Yacht Club, 21

Madura Strait, 213-4, 226
Malabar Coast, 154
Malabar Hotel, Cochin, 171
Malacca Strait, 186-7, 197
Malaya, (see also Penang and Ipoh), 177, 191-7, 229
   Home Guard, 192, 194-5, 216
Malta, 69-70
   ashore at, 72-3
   Church's influence in, 73
   Yacht Club, 71
Mannar, Gulf of, 173
Marsden, Colonel and Mrs., 166
Massawa, 114
Meda Flores (inn), 40
Melbourne, 301-2
Meridian of Greenwich, crossing of, 71
Mocha, 114
Monsoon, North-east, 76, 89, 142, 172
   periods and characteristics of, 120, 177
Monsoon, South-west, 122, 172, 180
   periods and characteristics of, 120-1, 177
Mooney, 'Doc', R.N., 200
Moplah Bay, 163

Mukalla, 129, 133, 138
   ashore at, 131-2
   Fatima and Farid, 132
   Sultan of, 132

Nancowry, 187
Nasser, President, 218
Nehru, Pandit, 155
Nelson (N.Z.), 313-4, 322
Nepal, 192
New Zealand, 25
   Bell Bird, 310-11
   D'Urville Island, 25
   passage to, 309-22
   'Welfare State', 109
Nicobar Islands, 187
North Island, W. Australia, 284
North-west Cape, Australia, 226, 229, 233
Northampton, W. Australia, 250-1, 278

Orion, s.s., 97

Penang, 177, 192, 195
   passage to, 179-190
   Roger, 191, 195
Perim Island, 112-5, 116, 117
   Mike, 115-19
Perth, 250, 271, 283, 289
   ashore at, 294-300
Petersen (lone voyager), 114-5
Pigeon Island, 284
Pilot Book, 30, 39, 104, 107, 112, 117,

129, 155, 166, 169-70, 185, 204, 233-4, 299, 311, 314
Port Fuad, 91
Port Philip, 301
Port Said, 56, 94, 97
   ashore at, 89-93
   Cercle Nautique, 90-3
   Emile, 91
   MacGregor, 90-91
   passage to, 85-89
Port Suez, 98
   interrogation at, 99-100
Portuguese Cafe, Algiers, 54
Prongs, The, 140
Puket Island, 187
Pulo Bras, 186

Raffles Light, 197
Ramadan, fast of, 131
Ras Engels, 69
Ras Gharib, 100
Ras Kodar, 132
Red Sea passage, 100-12
Refuge Cove, 301
   ashore at, 303-9
   passage to, 299-303
Rhio Strait and Archipelago, 204-5
Rio Mandovi, 154
Roaring Forties, 234, 299-300
Robeson, Paul, 65
Robinson, W. A. (lone voyager), 104, 129

Rondo, 186
Rottnest Island, 241-2
Royal Colombo Yacht Club, 173, 175-83
Royal Freshwater Bay Yacht Club, Fremantle, 289, 290
Royal Singapore Yacht Club, 200, 202

St. Francis Xavier, 158
Seymour-Williams, Colonel, 140
Shark Bay, 233, 245, 254, 278
Shelldrake, s.s., 26
Siam, 187-8, 223
Simpson, Bill, 191
Singapore, 213
   ashore at, 200-3
   Naval Base Sailing Club, 200
   passage to, 197-200
   Victoria, 202-3
Sirte, Gulf of, 76
Sisters Islands, 198
Skerki Channel, 69
Slocum, Joshua (lone voyager), 21
Socotra Island, 129, 132
Solent, The, 25
Spain, 40-1
Speedwell, yacht, 201-2
Spender, Stephen, 128
Sport Nautique, Algiers, 54
Steel Designer, s.s., 110
Stora, Gulf of, 60
Stornaway, Petersen's yacht, 114

INDEX

Strange, Albert, 114
Suez Canal, passage through, 94-9
Suez Canal Company, 91-4
Suez, Gulf of, 99-100
Sultan Shoal, 197
Sumatra, 177, 186
Sumatra Squalls, 197-200
Surabaya, 208-14, 226, 229-30
    Dutch hospitality at, 215-19
    Kitty and Fien, 215-19, 226, 228, 255
    passage to, 212-4
Swan River, 294
Sydney, 201

Taj Mahal Hotel, Bombay, 142
Tasman Bay, N.Z., 313
Tasman Sea, 303
Tasmania, 301
Teignmouth, Devon, 23-4
Templer, General (now Field Marshal) Sir Gerald, 195
Tenez, 48
Tensing, Sherpa, 226
Timor, 200-1, 207
Timsah, Lake, 96
Tobruk, 85-6

Tornus, s.s., 152
Trade Winds, South-east, 204, 226-8
Trimmerwheel, launch, 282-88
    Fred, 282-3
    Earle, 284
Tunis, Bay of, 69

Ushant, 25, 26
Venduruthy, I.N.S., 172
Vieux Port, Algiers, 50

Water distillation, 237-41
Western Ghats, 154
Westport, 315-320
Wilson's Promontory, 301-9
World Within World, 128

Yacht Club,
    Algiers, 54-5, 61
    Bône, 63, 66
    Lymington, 21
    Malta, 71
    Royal Colombo, 173, 175-6
    Royal Freshwater Bay, Fremantle, 289, 291
    Royal Singapore, 200, 202
Yemeni tribesmen, 114

Sheila II broke her moorings off Devonport Yacht Club in Auckland, New Zealand in 1984 and was badly holed on the rocks at Hobson Bay. She remains in that state of disrepair. The Albert Strange Association is in contact with her owner. At the time of this publication, *Sheila II* is for sale and it is hoped she will sail again.